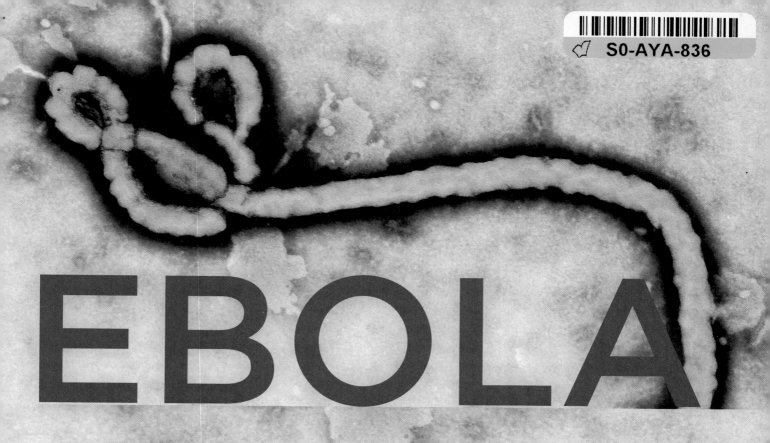

EBOLA

AN EMERGING INFECTIOUS DISEASE CASE STUDY

GEORGE T. EALY, MS, MD, PhD
Professor of Microbiology
Keiser University
Jacksonville, Florida

CAROLYN A. DEHLINGER, MAS, MS
Professor of Biological Sciences
Keiser University
Jacksonville, Florida

JONES & BARTLETT
LEARNING

World Headquarters
Jones & Bartlett Learning
5 Wall Street
Burlington, MA 01803
978-443-5000
info@jblearning.com
www.jblearning.com

Jones & Bartlett Learning books and products are available through most bookstores and online booksellers. To contact Jones & Bartlett Learning directly, call 800-832-0034, fax 978-443-8000, or visit our website, www.jblearning.com.

Substantial discounts on bulk quantities of Jones & Bartlett Learning publications are available to corporations, professional associations, and other qualified organizations. For details and specific discount information, contact the special sales department at Jones & Bartlett Learning via the above contact information or send an email to specialsales@jblearning.com.

Production Credits

Chief Executive Officer: Ty Field
President: James Homer
Chief Product Officer: Eduardo Moura
VP, Executive Publisher: David D. Cella
Senior Acquisitions Editor: Matthew Kane
Senior Development Editor: Nancy Hoffmann
Associate Editor: Audrey Schwinn
Associate Director of Production: Julie C. Bolduc
Production Assistant: Brooke Appe
Marketing Manager: Lindsay White

Manufacturing and Inventory Control Supervisor: Amy Bacus
Composition: Integra Software Services Pvt. Ltd.
Cover/Interior Design: Kristin E. Parker
Rights & Media Manager: Joanna Lundeen
Rights & Media Research Coordinator: Abigail Reip
Media Development Editor: Shannon Sheehan
Cover Image: © Frederick A. Murphy/CDC
Printing and Binding: RR Donnelley
Cover Printing: RR Donnelley

Library of Congress Cataloging-in-Publication Data

Ealy, George, 1950–, author.
  Ebola : an emerging infectious disease case study / by George Ealy and Carolyn Dehlinger.
    p. ; cm.
  Includes bibliographical references and index.
  ISBN 978-1-284-08778-9 (pbk. : alk. paper)
  I. Dehlinger, Carolyn, author.   II. Title.
  [DNLM: 1. Communicable Diseases, Emerging—prevention & control.   2. Hemorrhagic Fever,
Ebola—epidemiology.   3. Ebolavirus.   4. Epidemiologic Methods. WC 534]
  RC140.5
  614.5'7—dc23
                                                                      2015018545
6048

Printed in the United States of America
19 18 17 16 15   10 9 8 7 6 5 4 3 2 1

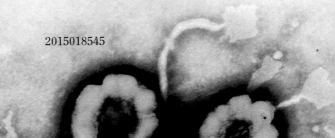

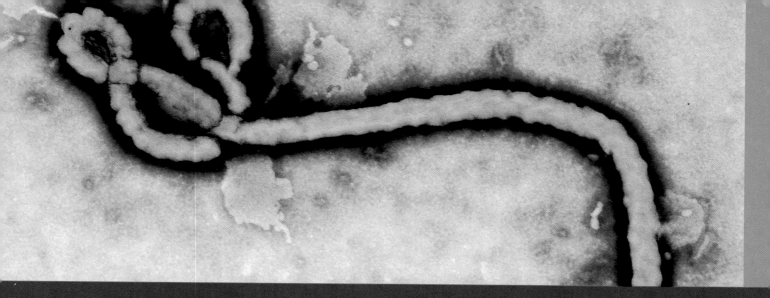

Dedication

This text is dedicated to the victims and survivors of Ebola virus disease, and to all those who fight against it.

George T. Ealy
Carolyn A. Dehlinger

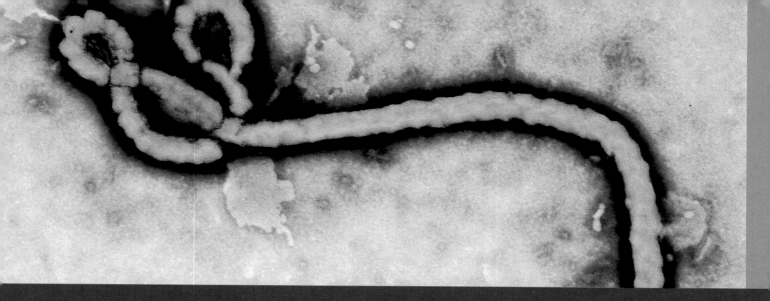

Brief Table of Contents

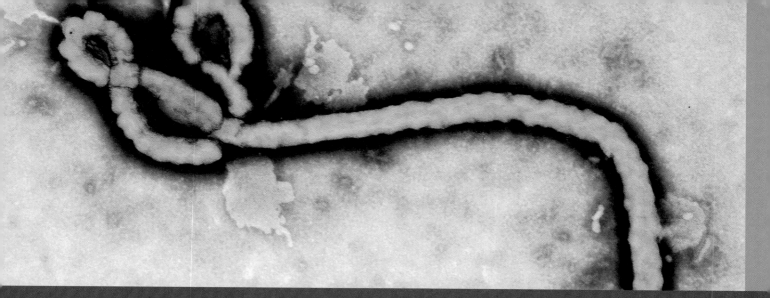

Table of Contents

CHAPTER 6   The Future of Ebola and Other Emerging
Diseases   **105**

Carolyn Dehlinger

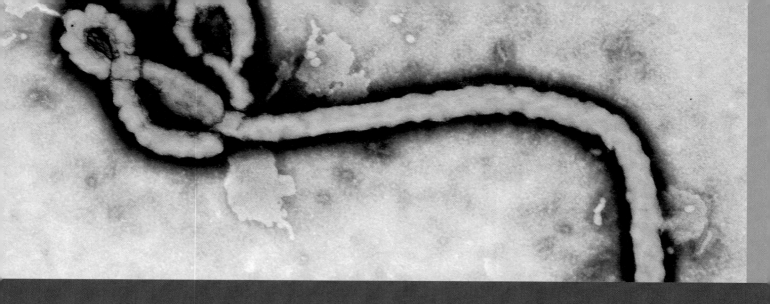

Foreword

This is not a traditional textbook about the Ebola virus. Instead, it is a discussion of Ebola in the context of the most recent outbreak, written from the perspectives of two professionals in the medical and science fields.

Much of what we know about Ebola has come from the media, and this text will reference many of the topics and events that are part of that collective understanding. But unlike what we have heard or read about in the news, *Ebola: An Emerging Infectious Disease Case Study* provides a firm foundation for academic discussions within microbiology, nursing, and public health programs. Informative and unique, this text should be a springboard to wider conversations about infectious diseases in the modern age and how we respond to them.

Ebola is analyzed here from a variety of perspectives, including the historical context of the 2014 outbreak, the biological makeup of the virus, and the associated diagnostic tests and drug treatments. Political and social angles are also explored in coverage of epidemiology, the role of the Centers for Disease Control and Prevention and World Health Organization, and the response of the governments directly affected by the virus.

Ebola: An Emerging Infectious Disease Case Study will provide answers for the curious science or healthcare student, but its real strength lies in the questions it raises and the subsequent discussions that it will elicit. How is Ebola similar to or different from other emerging infectious diseases? Why was this most recent outbreak more widespread than previous outbreaks of the virus? How did organizations and governments handle the response? What are the implications of deciding whether or not to quarantine? What can we expect from Ebola or other infectious diseases in the future?

In the space between current events and history sits *Ebola: An Emerging Infectious Disease Case Study*. The conversation has started…what will you contribute to it?

Matthew Kane
Senior Acquisitions Editor
Jones & Bartlett Learning

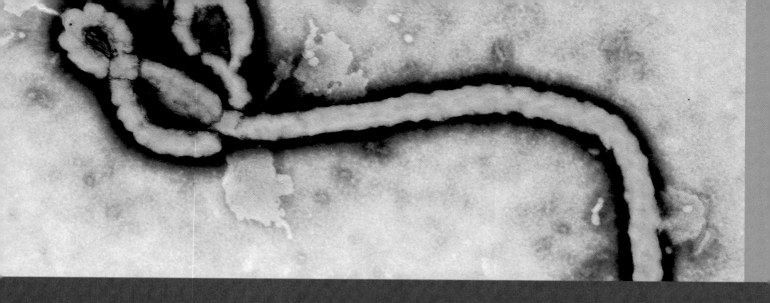

Preface

During 2014, an illness once confined to remote areas of Africa erupted from a single case into an outbreak that, by the year's end, had infected more than 22,000 people and killed nearly 9,000. To the public, this disease seemed like no other: It conjured thoughts of isolation, quarantine, and death from massive bleeding, both internal and external. The outbreak of Ebola virus disease created some tension due to the differing responses of international aid agencies, such as the World Health Organization (WHO), Doctors Without Borders/Médecins Sans Frontières (MSF), and the Centers for Disease Control and Prevention (CDC).

Some American citizens questioned the ability of the CDC to contain the illness after two nurses became infected following their treatment of an Ebola patient in Dallas. In 2014, *Time* selected the Ebola fighters as their collective Person of the Year. As direct care providers, these healthcare workers risked and often lost their lives in a valiant effort to contain the Ebola virus epidemic. Despite their efforts, the virus spread into several countries of West Africa that shared borders, causing a widespread epidemic. Such was the effect of Ebola: a disease for which at present there is no cure or approved vaccine, and for which treatment requires patient isolation and supportive therapy such as intravenous fluids and blood transfusions.

With the continuing increase in global communication and travel, we have truly become residents of a global community. The threat of emerging infectious diseases such as Ebola on an epidemic and even pandemic scale is not only a growing concern for all, but also a growing area of opportunity for research, collaboration, and exploration. This text takes the Ebola virus as a case study in emerging infectious disease and uses it as a lens to explore concepts such as biology, epidemiology, biotechnology, and global response. *Ebola: An Emerging Infectious Disease Case Study* was prepared with a wide audience in mind, but specifically students beginning to investigate their global community and the science inside it. It was envisioned to provide more depth to undergraduate courses in the biological sciences but could easily be of use to students of public health, health sciences, or anyone simply interested in learning more about the story of Ebola.

Another unique aspect of this text is its position in the greater Ebola crisis timeline. We are able to examine the outbreak within the context of all that transpired between early 2014 and the first half of 2015. However, as the outbreak further develops and even as new emerging infectious diseases surface, this case study work will remain static. Any reader of this text must become an active participant in the case study and should continue to gather more information about what has transpired. The CDC, WHO, MSF, and The Lancet's Ebola Resource Center are all great resources to begin a search for news on the latest developments.

What follows is neither the official scientific guide to the Ebola virus nor the final authority on patient care for its treatment. Rather, it is a reflective study of an emergent infectious disease that quickly captured the world's attention and came to dominate the 24-hour news cycle. At the intersection of science, public health, and media was Ebola, and this work is intended to start the conversation about what happened and its potential future implications.

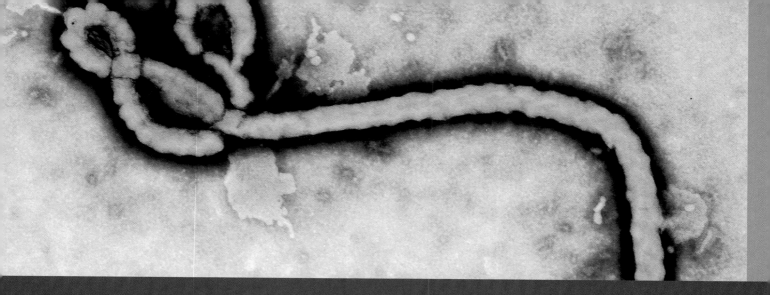

Acknowledgements

We would like to acknowledge the talented individuals at Jones & Bartlett Learning for providing support and guidance throughout the creation of this text. Senior Acquisitions Editor Matthew Kane shared his outstanding leadership and vision throughout this project. Senior Development Editor Nancy Hoffmann has offered irreplaceable advice and assistance. Audrey Schwinn, Associate Editor, remained patient and helpful while organizing the many details of the project. We are grateful to the production team members including Brooke Appe, Abby Reip, and Shannon Sheehan. We appreciate the permissions granted on behalf of the authors, editors, and publishers of all copyrighted material.

Dr. Ealy and Ms. Dehlinger are pleased to acknowledge the assistance of all the faculty reviewers who participated in this project. The knowledgeable contributions provided by these scholars helped to remove erroneous remarks and included useful elaboration on material. Their suggestions collectively improved the content and flow of this text and helped develop a student-centered presentation.

Lastly, we must thank our friends and family, especially our spouses, Roxanne Ealy and Charles Dehlinger, for generously donating the borrowed time we took in writing these pages. Dr. Ealy thanks his brother, W. Carl Ealy, and canine daughter, Gianna. He would also like to acknowledge in loving memory, his parents, Garnice Knight and Thomas Clark Ealy, who encouraged him in his early years to continue writing and to never stop learning. Ms. Dehlinger also thanks her parents, David and Virginia Baker, and sister, Jaime Flajole.

George T. Ealy
Carolyn A. Dehlinger

The authors and Publisher would like to thank the following individuals for serving as reviewers of the manuscript.

Robert J. Boylan, PhD, MS
New York University

Brian David Byrd, PhD, MSPH
Western Carolina University

Chad A. Corcoran, PhD
Le Moyne College

Eric DeAngelo, MS
Lehigh Carbon Community College

Joel Gaikwad, PhD, MS
Oral Roberts University

Diane Y. John, PhD, ARNP, FNP-BC
Frontier Nursing University

Ragan Johnson, DNP, APRN-BC
University of Tennessee Health Science Center

Reena Lamichhane-Khadka, PhD
Saint Mary's College

Pamela Monaco, MD
Molloy College

Mark S. Parcells, PhD
University of Delaware

Nancy Peifer, RN, PhD, MSN
Palm Beach State College

Bradley Skillman, MA
Florida State University

Ekaterina Vorotnikova, PhD
University of Massachusetts, Lowell

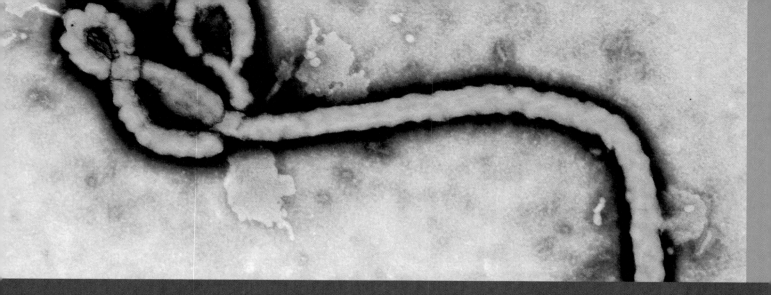

About the Authors

George Ealy, MD, PhD, completed his undergraduate education at Davidson College (BS), continued at the University of North Carolina at Charlotte to earn a Master of Science, and received his Doctor of Philosophy in Microbiology and Immunology in 1980 from the University of Louisville. He continued his postgraduate education there, earning the Doctor of Medicine in 1987.

A noted scholar of the composer Ludwig van Beethoven, Dr. Ealy published a seminal article in 1994 concerning the composer's famous hearing loss, which definitively established from primary source materials that Beethoven was able to hear throughout his life by using early technology for sound amplification (ear trumpets and various resonating devices at the piano). The article, "Of Ear Trumpets and Resonance Plate: Early Hearing Aids and Beethoven's Hearing Perception," was published by *19th-Century Music* (University of California Press), a renowned journal of musicology, and has been cited hundreds of times in various books and research articles. Dr. Ealy has also written several articles concerning the medical history of composers including Beethoven and Mozart.

Dr. Ealy is currently professor of microbiology at Keiser University (Jacksonville, FL) and instructs nursing and biomedical science students in clinical microbiology, for which he has received numerous excellence-in-teaching awards. He is also the author of a forthcoming microbiology textbook. In 2013, Dr. Ealy received the highest award given by the Commonwealth of Kentucky when he was commissioned by the governor as a member of the Order of Kentucky Colonels. He is also an active member of the Duval County Medical Society and the Florida Medical Association.

His lovely and loving wife, Roxanne Remondelli-Ealy, RN, BSN has inspired much of his writing, acting as both a sounding board and *de facto* editor for many of the clinical aspects related to microbiology. The other woman in his life is his beloved Italian greyhound, Gianna, who patiently stayed by his side during much of the writing for this text.

Carolyn A. Dehlinger is a professor of biological sciences at Keiser University in Jacksonville, Florida. She received her Master of Applied Science in Technology Management from the University of Denver and her Master of Science in Biology from Mississippi State University. She completed her undergraduate degree at the University of Florida. She teaches students in the Biomedical Sciences and Biotechnology degree programs in addition to students enrolled in non-majors biology courses. At Keiser, Carolyn played an integral role in the development of the Biomedical Sciences degree curriculum and has served on multiple National Science Foundation grant projects aimed at delivering biotechnology education to both students and educators. Carolyn serves as a subject matter expert with the American Council on Education in Washington, DC. In this role, she evaluates curricula and determines their eligibility for accreditation. She is a proud member of the Honorable Order of Kentucky Colonels, an honor bestowed by the governor of Kentucky for community service. Carolyn is a practitioner and registered teacher of yoga and enjoys spending her free time with her husband Charles; their two golden retrievers, Alice and Ivan; and their calico cat, Chloe. She is also the author of Jones & Bartlett Learning's textbook, *Molecular Biotechnology*.

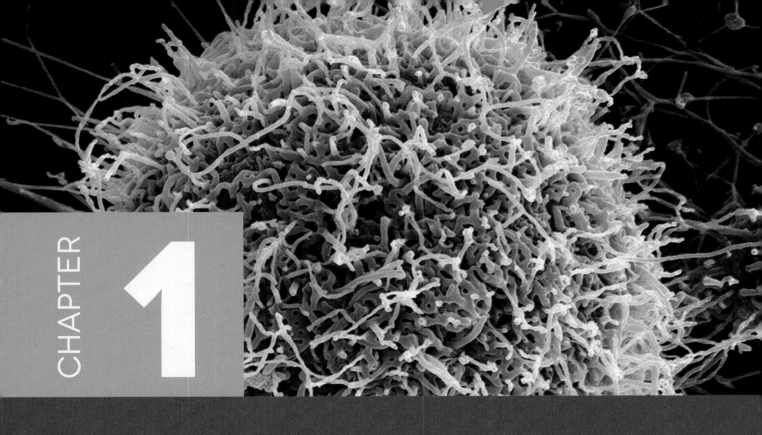

CHAPTER 1

The Natural History of Ebola Virus

George Ealy

This chapter addresses the increasingly important global healthcare problem of emerging infectious diseases as evidenced by the recent outbreak/epidemic of Ebola virus. Previous outbreaks of Ebola virus disease and their successful containment are examined. Epidemiological methods for examining an emerging infectious disease are also presented. Methods used in molecular biotechnology to identify and contain the spread of a microorganism once an outbreak has been determined are also presented together with strategies for assessing their effectiveness. Finally, the text explores the development of vaccines and antiviral agents for the Ebola virus and the possible use of the virus by bioterrorists.

■ QUESTIONS TO CONSIDER

Some questions to consider as you read this chapter include:

1. What is the difference between an outbreak and an epidemic? Is the Ebola virus disease a pandemic?

2. Until 2014, Ebola virus disease had occurred in countries that are in the sub-Saharan desert regions of Africa. Discuss ways in which the virus moved from the equatorial regions of Africa to the west coast.

3. What is the meaning of antigenic drift and how might it relate to the Ebola virus?

4. Explain the chain of infection and how the links of the chain are related to the occurrence of a disease.

5. Explain the difference between direct and indirect contact and give several examples of each method of transmission.

6. Why is determining the index case important in epidemiology?

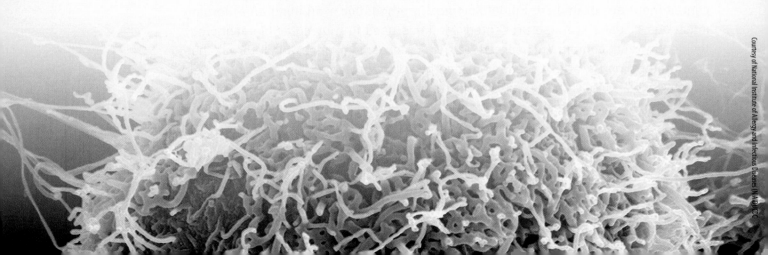

►► Factors Influencing the Emergence of the Ebola Virus Outbreak

Ebola virus disease is one of a group of illnesses that are regarded as emerging infectious diseases (EIDs). Two types of EIDs are recognized. One group is comprised of completely new illnesses that appear within a population without precedent, such as the human immunodeficiency virus (HIV) and its end-stage condition called acquired immuno-deficiency syndrome (AIDS). A second type includes diseases that were once common, but have decreased to the point of almost nonexistence followed by an increase in the number of new cases. The Centers for Disease Control and Prevention (CDC) considers an EID to be an infectious illness that has increased in numbers of newly diagnosed cases within the past 20 years. This includes not only new diseases but also recognized illnesses that are seeing an increase in numbers of cases for a variety of reasons. A partial list of EIDs is provided in **TABLE 1.1** .

A number of factors played a role in the huge outbreak of Ebola virus in 2014. Previous outbreaks had always been small and occurred in isolated villages, causing the illness to be self-contained. Civil unrest and war, combined with porous borders, caused people to move in large numbers from area to area, facilitating the disease's spread.

TABLE 1.1	Partial List of Recent Emerging and Reemerging Infectious Diseases		
Microorganism	**Disease**	**Year**	**Reservoir**
Dengue virus	Dengue "breakbone fever"	1800s	Humans
Chikungunya virus	Dengue-like illness	1952	Humans
Nipah virus	Encephalitis	1999	Pigs, bats
Coronavirus	Severe acute respiratory syndrome (SARS)	2003	Bats?
Middle East respiratory syndrome coronavirus (MERS-CoV)	Severe respiratory infection	2012	Bats?
Poliovirus (Wild Type 1; originating in Saudi Arabia)	Poliomyelitis	2013	Humans
H7N9 influenza	Avian influenza	2013	Poultry
Measles virus	Measles	2013	Humans

Social customs such as the reliance on healers, traditional funeral practices, and the practice of eating wild animal meat also contributed to the outbreak. Increasing contact between humans and the suspected animal reservoir (fruit bats) led to the phenomenon of zoonotic spillover—also a contributing factor.

Climate changes affect the growth of fruit trees from which the bats feed. During periods of dry weather less fruit is produced, causing bats to migrate to remote areas with more fruit growth, which leads to less contact between humans and fruit bats. However, fruit trees flourish when rain is plentiful, creating increased opportunities for fruit bat–human interaction. Bat fecal matter may end up on fruit that humans consume or humans may hunt bats for bushmeat (wild meat eaten in parts of eastern Africa).

The Role of Mutation and Evolution

A number of factors favor the emergence of disease. Microbes—like all living organisms—undergo changes in their genetic composition. These changes are called **mutations** and are the driving force for evolution. Evolution represents an adaptation to pressures that are placed on living organisms. The introduction of antibiotics acts as a selective pressure on bacteria: They either evolve strategies to resist the killing effect of the antibiotic or they die. Bacteria are very adept at overcoming the effects of antibiotics. In nature, antibiotics are produced by bacteria or fungi as a means of eliminating competitors for nutrition and energy. The number of microorganisms that are exposed to an antibiotic in a natural setting is limited compared to the number that become exposed to an antibiotic from pharmaceutical manufacture and marketing. The global exposure of bacteria to antibiotics since the 1950s became a driving force for microbial evolution, resulting in adaptations that have caused many antibiotics to no longer be effective. This phenomenon, called **antibiotic resistance** (see **FIGURE 1.1**), is a factor in the growth of emerging bacterial diseases.

Microorganisms can also develop mutations that are unrelated to antibiotics. This is especially true for viruses. The influenza virus undergoes yearly mutations that are referred to as **antigenic drift**. Naturally occurring mutations are common in the influenza virus and are the basis for antigenic drift. These are changes in the viral genes that produce different surface proteins that act as docking devices for the virus so that it can enter a target respiratory cell. Antigenic drift is responsible for the occasional failure of the seasonal influenza vaccine, because the new surface proteins are not recognized by the body's immune system.

Mutations are also responsible for the emergence of viral species. The Ebola virus has five known species, four of which cause human disease. The current epidemic is caused by **Zaire ebolavirus**, which we know based on genetic sequencing. Although the source for the Ebola virus remains obscure, it is possible that mutations in a very similar virus (Marburg) were responsible for the appearance of Ebola. Similarly, mutations that occurred within the ribonucleic acid (RNA) genome of the Ebola virus are most likely responsible for the appearance of new strains.

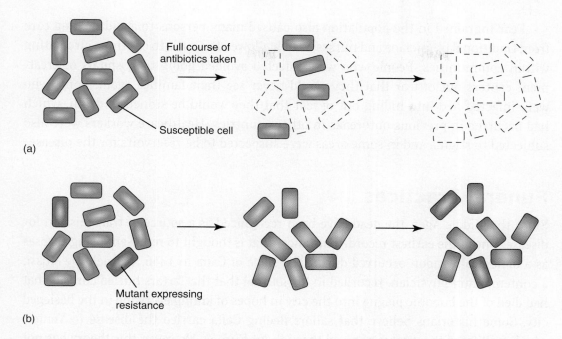

(a)

Full course of
antibiotics taken

Susceptible cell

(b)

Mutant expressing
resistance

■ **FIGURE 1.1** The possible outcomes of antibiotic treatment: (a) Ideally, with a complete course of antibiotics, all pathogens will be destroyed. (b) If there are some resistant cells in the infecting population, they will survive and grow without any competition.

Demographics and Social Conditions

Population growth and human migration have played a role in the emergence of historical epidemics such as the bubonic plague and yellow fever. Similar factors seem likely in the current Ebola epidemic. Human migration in West Africa exceeds that of the rest of the world by a seven-fold factor. This is driven in part by decades of civil unrest and conflict as well as the search for better economic living conditions. Continuous armed conflict occurred in the West African countries of Sierra Leone, Guinea, Liberia, and the Ivory Coast (Côte d'Ivoire) from 1999 until 2004. This degree of armed conflict was responsible for some of the migrations that occurred within the region. According to the Public Library of Science (PLOS), the degree of rural-to-urban migration has been immense in the decades between 1960 and 2013 in Guinea, approaching 250%, and almost 150% in both Sierra Leone and Liberia.

An additional factor favoring emergence of the virus is the tremendous population growth that has also occurred in this area of the continent. Population densities have increased in Liberia, Guinea, and Sierra Leone by as much as 200% in the same time period between 1960 and 2013. The 2014 outbreak was the first time that cities were the focus of Ebola infection; all previous outbreaks occurred only in rural settings. World Health Organization (WHO) data show the number of physicians per 1,000 people to range from 0.1 in Guinea to 0.02 in Sierra Leone (compared to 2.45 per 1,000 in the United States and 2.79 in the United Kingdom). The increase in urban populations outpaced the ability of healthcare resources to provide care. This in part contributes to a high reliance on nontraditional medical care in these countries.

Fear ingrained in the population also caused many persons to avoid seeking care from traditional physicians and nurses, which allowed the virus to continue circulating within communities. People who were infected avoided going to hospitals or treatment centers out of fear that they would never see their families again. Some who were infected fled into hiding out of fear that they would be stoned to death, which had occurred in previous outbreaks in other countries. Healthcare workers were also subjected to stigma and in some areas were suspected to be reservoirs for the disease.

Funeral Practices

Since the Middle Ages, the dead have been recognized as a source of transmission for disease. One of the earliest recorded instances that is thought to represent using corpses as a biological weapon occurred during the siege of Caffa in 1346. Gabriele De'Mussi, a contemporary physician, recorded in his journal that the Tartars hurled corpses that had died of the bubonic plague into the city in hopes of killing everyone in the besieged city. Some historians believe that sailors fleeing Caffa carried the disease to Venice, which facilitated the disease's spread throughout Europe. However, this theory has not been confirmed.

In Guinea, more than 60% of Ebola infections resulted from burial practices that involved washing the body of the deceased. The tradition of mourners touching the deceased's body also favors the continued transmission of the disease.

Zoonotic Spillover

The Ebola virus is believed to exist as a **zoonosis**, although an animal reservoir remains undetermined. Zoonosis refers to a disease that normally infects animals and can be transmitted to humans. Several species of fruit bats of the *Pteropodidae* family are widely believed to be the reservoir for the Ebola virus, but this association has not been proven, in part because the virus has never been cultured from fruit bats—a necessary requirement to prove their involvement. A recent investigation published by *EMBO Molecular Medicine* in 2014 suggests that another type of bat, *Mops condylurus* (commonly known as the Angolan free-tailed bat), may be a possible candidate for the long-sought-after reservoir. Although not conclusive, epidemiologists have observed a possible association between children who eat Angolan bats and the index case for the current Ebola outbreak. From interviewing people in the village where the index case occurred, epidemiologists found that children hunted and ate Angolan bats that were living in a tree close to the home of the index case.

Despite this finding, many epidemiologists strongly believe that the straw-colored fruit bat (*Eidolon helvum*)—which has a migratory range of up to 2,500 kilometers—could be responsible for the spread of the virus from Central to West Africa. **Zoonotic spillover** refers to the movement of a disease from animals to humans. In the case of Ebola, it may have occurred from human contact with either infected bats or animals that had become infected by bats. Several types of fruit bats are shown in FIGURE 1.2, while the Angolan free-tailed bat is depicted in FIGURE 1.3.

(a)

(b)

(c)

(d)

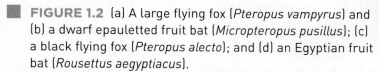

FIGURE 1.2 (a) A large flying fox (*Pteropus vampyrus*) and (b) a dwarf epauletted fruit bat (*Micropteropus pusillus*); (c) a black flying fox (*Pteropus alecto*); and (d) an Egyptian fruit bat (*Rousettus aegyptiacus*).

Although nonhuman primates are *not* considered to be the primary reservoir for the virus, they can become infected with the virus. Infection of nonhuman primates probably occurs from eating fruit contaminated with either bat feces or saliva. Humans who eat **bushmeat** infected with Ebola virus may become infected themselves. Bushmeat refers to the flesh of wild animals such as monkeys, gorillas, and bats. Bushmeat is a part of the staple diet in rural West Africa and parts of Asia. Other than through handling or consuming infected meat, how infection occurs in these settings is unknown at the present time.

Climatological Factors

Weather may have a direct effect on the occurrence of Ebola outbreaks; meteorological satellite analysis from 1981 to 2000 demonstrated a correlation between Ebola outbreaks and sudden alterations in climate from dry to wet conditions. Fruit trees in periods of little rainfall produce less fruit. When heavy rains occur, trees bear more fruit, leading to both bats and nonhuman primates gathering in the same areas for food. Such close association provides opportunities for the virus to jump between species. As noted earlier, humans can become infected by handling or eating infected animals.

WHO has also noted a correlation between rising global temperatures and an increase in other types of infectious disease. Rising temperatures favor the spread of diseases that are transmitted by vectors such as insects that thrive in warmer temperatures. Emerging diseases such as a variety of equine encephalitides, West Nile encephalitis, and dengue fever are spread by mosquitoes, which increase in numbers with higher temperatures.

Environmental Changes

For a number of years, environmentalists have warned that the extensive deforestation practices in Africa and South America would have unforeseen impacts on climate and health. Deforestation in Central and West Africa has been mainly the enterprise of international mining and timber companies. As the forested areas of West Africa disappeared, the buffer zone between zoonotic animals and humans correspondingly diminished, thus providing opportunities for the occurrence of virus spillover. An example of deforestation is shown in **FIGURE 1.4**.

Public Health Availability and Infrastructure

Inadequate public health structure is also a factor in the spread of infectious diseases. Access to public health care is limited or absent in many African countries. A typical hospital in Guinea, shown in **FIGURE 1.5**, illustrates the limited facilities available for treatment. Liberia, which has a population of 4 million, had fewer than 200 physicians prior to the Ebola outbreak. Each of the three

FIGURE 1.3 An Angolan free-tailed bat (*Mops condylurus*).

Bernard DUPONT/Flickr https://creativecommons.org/licenses/by-sa/2.0/

FIGURE 1.4 Deforestation can increase the likelihood of contact between humans and zoonotic animals.

© Photodisc

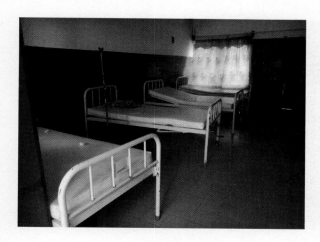

FIGURE 1.5 Photographed during Guinea's 2014 Ebola outbreak, this image depicts an interior view, inside the Infectious Diseases ward at Donka Hospital located in the country's capital city of Conakry. Here you can see the minimalistic layout of the patients' quarters.

Courtesy of Dr. Heidi Soeters/CDC

most affected countries ranks at nearly the bottom of the United Nations Human Development Index. This is a composite index that reflects income, education, and human life expectancies that was formulated by Pakistani economist Mahbub ul Haq in 1990 as a means of ranking countries' quality of life. It is used by the United Nations as an assessment tool for identifying areas of improvement. Norway has the highest ranking, with the United States ranking fifth.

TABLE 1.2 shows a number of emerging infectious diseases and some of the changes that have prompted their emergence.

TABLE 1.2	Emerging Infectious Diseases and Changes in Environment, Host, or Organism That Have Prompted Their Emergence
Anthrax	Bioterrorism

Arenaviruses
  Junin virus (Argentine
    hemorrhagic fever [HF])
  Machupo virus (Bolivian HF)
  Guanarito virus (Venezuelan HF)
	Changes in agriculture allowing closer contact with infected rodents

Borrelia burgdorferi (Lyme disease)	Increased deer population, increased human contact with ticks in nature
Cholera	El Niño climate change, international travel, shipment of food
Cryptosporidium parvum	Contamination of municipal water supplies, increases in immunocompromised populations
Cyclospora cayetanensis	International shipment of raspberries
Dengue	Increased global travel, urbanization, increase in mosquito reservoir

(continues)

TABLE 1.2	Emerging Infectious Diseases and Changes in Environment, Host, or Organism That Have Prompted Their Emergence (*Continued*)
Escherichia coli 0157:H7 (enterohemorrhagic *E. coli*)	Growth-centralized agriculture promoting cross-contamination, global distribution of foods
Filoviridae species Ebola-Marburg virus	Increased contact between infected primates and man, nosocomial spread, importation of animals
Hantavirus Sin Nombre virus (HPS)	Climatic changes allowing mice expansion
HIV/AIDS (HTLV)	Changes in sexual behavior, urbanization, increase in illicit drug use, global shipment of blood products
Influenza	Integrated pig-duck agriculture, increase in global travel
Malaria	Growth and movement of human populations, declining use and effectiveness of insecticides, crowding
Monkeypox	Human contact with exotic animals, animal contact
Multidrug-resistant *Mycobacterium tuberculosis*	Misuse of antibiotics, crowding in prisons, slums, hospitals, etc., allowing transmission
Pfiesteria piscicida	Changes in agricultural practices leading to pollution of rivers and estuaries, overgrowth of dinoflagellates
Quinolone-resistant *Campylobacter*	Overuse and misuse of antibiotics in agriculture and clinical settings
Raccoon rabies	Shipment of infected raccoons
Rift valley fever	Dams, irrigation, climate change
Severe acute respiratory syndrome (SARS)	Human contact with exotic animals (civet cats), international travel

▶▶ History of Ebola Virus Outbreaks

Marburg Virus: A Precursor for Ebola?

Ebola virus is closely related to the Marburg virus that was first isolated in 1967 in Marburg, Germany. Outbreaks of **hemorrhagic fever** occurred nearly simultaneously among laboratory workers in the German cities of Marburg and Frankfurt and in Belgrade, Yugoslavia (now Serbia). In each laboratory, those infected were handling tissues of African green monkeys. African fruit bats are the reservoir host for the Marburg virus, which is the same animal presumed to be the reservoir for Ebola virus. The disease caused by both viruses is clinically similar.

Ebola Outbreaks: 1976–1979

The first known outbreak of Ebola virus occurred in June 1976 in Zaire (now known as the Democratic Republic of Congo). A photograph of the area near the start of the 1976 outbreak is shown in **FIGURE 1.6**. This outbreak was the most severe of all subsequent outbreaks (prior to the current one) in terms of numbers of deaths, with a mortality rate of 88%; the outbreak was spread by the use of contaminated and unsterilized needles that were being used by healthcare workers. A photograph of the hospital during the 1976 outbreak is shown in **FIGURE 1.7**.

FIGURE 1.6 In the background of this photograph of a Maridi, Sudan, dirt-road intersection are a number of thatched dwellings representing the Maridi Hospital, which was located in one of the areas in Sudan that was devastated by that region's 1976 Ebola virus outbreak.

Courtesy of Dr. Lyle Conrad/CDC

FIGURE 1.7 The Maridi Hospital in Maridi, Sudan, which was where the first Ebola patients were treated in 1976.

Courtesy of Dr. Lyle Conrad/CDC

The **index case** for the entire history of the Ebola virus disease has been traced to a patient who developed symptoms on September 1, 1976, after receiving treatment for malaria 5 days earlier at Yambuku Mission Hospital. Within a week, other patients receiving treatment at the same hospital developed symptoms of Ebola virus infection. All patients had received injections of chloroquine for the treatment of malaria. The symptoms of both the Ebola virus disease and malaria are the same during the first days of the illnesses. Later investigations revealed that healthcare personnel at the hospital were reusing the same needles for injection, without sterilization. This was the cause for transmission of the first outbreak; however, how the first patient initially became infected was never determined. The cause of the first outbreak was identified as a virus closely resembling, but not the same as, the Marburg virus of 1967. Because the virus was first identified in Zaire (Democratic Republic of Congo) in an area near the Ebola River, the virus was named Ebola Zaire.

A second outbreak occurred the same year in Sudan, a neighboring country, and the virus was named Ebola Sudan. How the second outbreak occurred in a different country was also never determined. Spread among those who became infected was related to close contacts within hospitals, and many hospital workers became infected. Both outbreaks were contained by quarantine and isolation.

Ebola Outbreaks: 1979–2014

After the initial outbreaks of 1976 were contained, the next significant outbreak occurred in Sudan during 1979 and was likewise contained. The disease did not reappear in Africa again until some 15 years later, in 1994. Between 1994 and 1996 five independent outbreaks occurred. During the years between the first outbreak in 1976 and its reappearance in 1994, the virus is believed to have been maintained by circulating in its natural reservoir host, most likely the African fruit bat.

The outbreaks between 1994 and 1996 provided the opportunity for intensive study of the disease by international teams of epidemiologists. During this time a number of facts concerning transmission were determined, such as the requirement for close contact with bodily fluids. Burial practices—particularly touching of the deceased—were also shown to be a major transmission factor.

After 1996, no outbreaks of Ebola virus occurred in Africa until 2000 when an outbreak lasting 1 year occurred in Uganda. During this outbreak caused by Ebola Sudan, 425 people became infected, of which more than 50% died. Epidemiologists determined that the major risk factors for acquiring the disease were related to attending funerals of the deceased, close contact with infected patients within a family group, and failure of healthcare workers to wear protective equipment.

From the year 2001 until 2014, limited outbreaks occurred in several countries of Central Africa, principally Uganda and the Democratic Republic of Congo. In each case, the outbreak was contained. During an outbreak in 2007 in Uganda, a new strain of Ebola was identified and named Bundibugyo for the district in which it was discovered. **TABLE 1.3** lists the known cases and outbreaks of Ebola.

TABLE 1.3	Known Cases and Outbreaks of Ebola Hemorrhagic Fever (EHF) in Chronological Order			
Year(s)*	Country	Ebola Subtype	Number of Cases (Mortality, %)	Situation
1976	Zaire (now DRC)	Ebola-Zaire	318 (88)	Spread by close personal contact and nosocomially by contaminated needles and syringes. First recognitions of EHF.
1976	Sudan	Ebola-Sudan	284 (53)	Spread through close personal contact nosocomially. Many medical care personnel were infected.
1976	England	Ebola-Sudan	1 (0)	Laboratory infection by needle-stick.
1977	Zaire	Ebola-Zaire	1 (100)	Noted retrospectively.
1979	Sudan	Ebola-Sudan	34 (65)	Recurrence at 1976 Sudan epidemic site.
1994	Gabon	Ebola-Zaire	52 (60)	In gold-mining camps deep in the rainforest. Initially thought to be yellow fever.
1995	Zaire	Ebola-Zaire	315 (81)	Index case worked in a forest adjoining the city. Spread through families and hospitals.
1996	Gabon	Ebola-Zaire	37 (57)	Chimpanzee found dead in forest was eaten; 19 cases were involved in the butchery; other cases in family members.
1996–1997[1]	Gabon	Ebola-Zaire	60 (74)	Index case was a hunter who≈lived in a forest camp. Spread by close contact with infected persons. A dead chimpanzee concurrently found in the forest was infected.

(*continues*)

TABLE 1.3	Known Cases and Outbreaks of Ebola Hemorrhagic Fever (EHF) in Chronological Order (*Continued*)			
Year(s)*	Country	Ebola Subtype	Number of Cases (Mortality, %)	Situation
1996	South Africa	Ebola-Zaire	2 (50)	A healthcare worker in the Gabon epidemic returned to South Africa; the nurse who took care of him also became infected and died.
2000–2001	Uganda	Ebola-Sudan	425 (53)	Risk factors were attending funerals of cases, being in the family of cases, and providing medical care without protection.
2001–2002	Gabon	Ebola-Zaire	65 (82)	Outbreak occurred on the border of Gabon and the Republic of Congo.
2001–2002	Republic of Congo	Ebola-Zaire	57 (75)	First Ebola infection case in the Republic of Congo
2002–2003	Republic of Congo	Ebola-Zaire	143 (89)	Outbreak occurred in the districts of Mbomo and Kéllé in Cuvette-Ouest Département.
2003	Republic of Congo	Ebola-Zaire	35 (83)	Outbreak occurred in Mbomo and Mbandza villages located in Mbomo district, Cuvette-Ouest Département.
2004	Sudan	Ebola-Sudan	17 (41)	Concurrent with an outbreak of measles; several suspect cases reclassified as measles.
2007	DRC	Ebola-Zaire	264 (71)	Outbreak occurred in Kasai Occidental Province.
2007–2008	Uganda	Ebola-Bundibugyo	131 (37)	First reported occurrence of a new strain.
2008–2009	DRC	Ebola-Zaire	32 (47)	Outbreak occurred in the Mweka and Luebo health zones of the Province of Kasai Occidental.

Year(s)*	Country	Ebola Subtype	Number of Cases (Mortality, %)	Situation
2011	Uganda	Ebola-Sudan	1 (100)	The Uganda Ministry of Health informed the public a patient with suspected Ebola Hemorrhagic fever died on May 6, 2011, in the Luweero district, Uganda. The quick diagnosis from a blood sample of Ebola virus was provided by the new CDC Viral Hemorrhagic Fever laboratory installed at UVRI.
2012	Uganda	Ebola-Sudan	11 (36)	Laboratory tests of blood samples were conducted by the UVRI and the CDC.
2012	DRC	Ebola-Bundibugyo	36 (36)	The outbreak in DRC had no epidemiologic link to the near contemporaneous Ebola outbreak in the Kibaale district of Uganda.
2012–2013	Uganda	Ebola-Sudan	6 (50)	CDC assisted the Ministry of Health in the epidemiologic and diagnostic aspects of the outbreak. Testing of samples by CDC's Viral Special Pathogens Branch occurred at UVRI in Entebbe.
2014	DRC	Ebola	66 (74)	The outbreak was unrelated to the outbreak of Ebola in West Africa.
2014–2015	Multiple countries	Ebola	25,626 (unknown)	At the time of writing, this outbreak is still ongoing. As of April 13, 2015, the CDC has reported 25,626 human cases including 10,619 deaths. The number of cases and deaths continues to evolve due to the ongoing investigation. Affected areas include all of Guinea, Liberia, and Sierra Leone and portions of Nigeria, Senegal, Spain, the United States, Mali, and the United Kingdom.

TABLE 1.3	Known Cases and Outbreaks of Ebola Hemorrhagic Fever (EHF) in Chronological Order (*Continued*) Outbreaks of Ebola-Reston Hemorrhagic Fever All cases were linked to animals in or from the Philippines (monkeys and pigs). No human illnesses have been recorded, but antibody seroconversions have occurred.		
Year(s)*	Country	Animal	Situation
1989	USA	Monkey	At quarantine facilities in Virginia and Pennsylvania, among monkeys imported from the Philippines. No human cases.
1990	USA	Monkey	At quarantine facilities in Virginia and Texas, in Philippine monkeys. Four humans developed antibodies asymptomatically.
1989–1990	Philippines	Monkey	High mortality among cynomolgus macaques. Three humans developed antibodies asymptomatically.
1992	Italy	Monkey	At quarantine facilities, among monkeys imported from the same export facility in the Philippines. No human cases.
1996	USA	Monkey	At quarantine facility in Texas, by Philippine monkeys. No human cases.
1996	Philippines	Monkey	At export facility in the Philippines. No human cases.
2008	Philippines	Pig	First known occurrence of Ebola-Reston in pigs. Six pig farm and slaughterhouse workers developed antibodies asymptomatically.

*Epidemics were concentrated in time. Those noted as occurring in 2 years were only several months long, but crossed over the 2 calendar years.
AIDS = acquired immunodeficiency syndrome; CDC = Centers for Disease Control and Prevention; DRC = Democratic Republic of Congo; HIV = human immunodeficiency virus; USA = United States of America; UVRI = Uganda Viral Research Institute.
Source: Modified from Centers for Disease Control and Prevention. (2015). Outbreaks chronology: Ebola virus disease. Retrieved from http://www.cdc.gov/vhf/ebola/outbreaks/history/chronology.html. Last modified April 13, 2015. Accessed April 14, 2015.

Ebola Outbreaks Outside of Africa

The United States experienced two very limited outbreaks of Ebola virus between 1989 and 1990 among laboratory workers. Macaque monkeys that were imported from the Philippines to Reston, Virginia, developed simian hemorrhagic fever. While investigating the cause of this outbreak, electron microscopy revealed the presence of filamentous-shaped virus particles that resembled Ebola Zaire. Several laboratory workers developed antibodies to the Ebola-like virus, which was identified as a new Ebola species and named Ebola-Reston (also known as Reston virus). This species of Ebola virus is not **pathogenic** for humans, and none of the workers developed any symptoms.

▶▶ Tracking the Index Case of 2014

Importance of the Index Case

The investigation of a disease outbreak is part of the field of epidemiology. Epidemiologists study when a disease begins, who is infected, how it is spread, and methods for its containment and prevention. Virtually all disease investigations are conducted retrospectively, because several cases must be diagnosed and reported to health departments before healthcare officials are aware that an outbreak has occurred. A period of time elapses before public health officers become aware that a problem exists. The goal of all epidemiological investigations is to identify as quickly as possible the very first person who became infected. As noted earlier, the first patient of an outbreak is referred to as the index case.

Determining the index case has several important functions in epidemiology, including shortening the time during which a disease spreads, aiding in identifying the source of an infectious disease, and possibly eliminating the source of infection. The faster that the index case can be identified, the shorter the time required for **contact tracing**. Contact tracing is basically medical detective work. Any person who has come into contact with the index case patient must be found and interviewed to determine their state of health. The longer it takes to find the index case, the more opportunities exist for transmission of an illness. Quick identification of the index case shortens the time for contact and can help to slow the transmission of an illness.

Finding the index case aids in identifying the source of a disease. Interviewing the first patient can reveal information regarding possible contacts with a disease reservoir. Information gathered from the index case patient may lead to identifying the source of an infectious agent, such as eating infected bushmeat.

How the Index Case for the Ebola Epidemic Was Identified

Determining the index case requires back-tracking medical records using a set of common symptoms as a unifying thread. In the case of the Ebola epidemic of 2014, the Ministry of Health of Guinea became aware on March 10 that a cluster of a then-mysterious disease with a high mortality rate was being reported by hospitals and public health officials in the prefectures of Guéckédou and Macenta. Two days later, Doctors Without Borders (DWB; or Médecins sans Frontières) in Guinea was also notified of the clustering. Both the Ministry of Health and DWB sent teams to the villages to investigate and collect samples from patients to determine the cause. The teams collected blood samples from a number of patients in the affected villages who shared the common symptoms of fever, vomiting, diarrhea, and bleeding. These teams also collected information regarding both demographics and clinical data at the time of sampling. Biotechnological methods were used to identify the causative agent as Ebola Zaire.

Epidemiological investigations of the path of transmission consisted of medical record reviews and interviews with families of infected patients, infected patients, and inhabitants of affected villages. Retrospective examinations established that the first infected patient was a 2-year-old in the village of Meliandou in the prefecture of Guéckédou who died on December 6, 2013. According to the CDC, the date of onset of symptoms for the index patient was December 2, 2013.

▶▶ Chain of Infection

A **chain of infection** is a system for analyzing an infectious disease in terms of links in a chain, where each link represents a crucial event necessary for the disease to exist. In order for an infectious disease to spread within a population, six interlinked elements must be present; interrupting any link in the chain will theoretically stop the infection. The links are: the infectious agent, a reservoir, a portal for exit, a mode of transmission, a portal for entry, and a susceptible host. The chain of infection is shown in FIGURE 1.8 . The Ebola epidemic is analyzed in terms of each element in the chain. Strategies for interrupting the chain of infection can be used to prevent future epidemics.

The Infectious Agent

Ebola virus, like other animal viruses, goes through a cycle of events in its lifecycle. The virus has to attach to a target cell(s), enter the cell, undergo synthesis of its genetic material and structural proteins, self-assemble its components, and then become released from the infected cell. The first step in the process offers a strategy for interrupting its lifecycle: Blocking attachment to the target cell(s) would prevent the later events and interrupt the chain of infection.

All Ebola virus species are RNA viruses surrounded by a protein shell that is further enclosed within a membranous structure called the envelope. Glycoprotein molecules extending from the envelope's surface function as receptors for viral entry into a susceptible cell. The structure of an Ebola virus is shown in FIGURE 1.9 . Vaccines cause the immune system to produce antibody molecules that bind to viral receptor proteins and prevent attachment.

Reservoir

A reservoir is the element in which an infectious agent perpetuates itself; it can be a human, an animal, or an aspect of the environment. TABLE 1.4 lists some common zoonotic infections and their reservoirs. In the case of Ebola, the reservoir remains unresolved. Finding the natural reservoir for Ebola could help in predicting where, when, and how often outbreaks might occur. Most scientists believe that African fruit bats are its natural reservoir. Circumstantial evidence supports the role of fruit bats as the natural

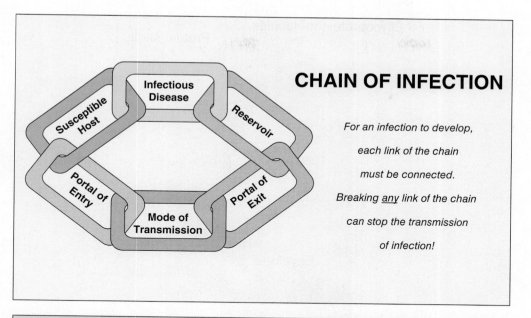

CHAIN OF INFECTION

For an infection to develop,

each link of the chain

must be connected.

Breaking <u>any</u> link of the chain

can stop the transmission

of infection!

Infectious Disease
Any microorganism that can cause a disease such as a bacterium, virus, parasite, or fungus. Reasons that the organism will cause an infection are virulence (ability to multiply and grow), invasiveness (ability to enter tissue), and pathogenicity (ability to cause disease).

Reservoir
The place where the microorganism resides, thrives, and reproduces, i.e., food, water, toilet seat, elevator buttons, human feces, respiratory secretions.

Portal of Exit
The place where the organism leaves the reservoir, such as the respiratory tract (nose, mouth), intestinal tract (rectum), urinary tract, or blood and other body fluids.

Mode of Transmission
The means by which an organism transfers from one carrier to another by either direct transmission (direct contact between infectious host and susceptible host) or indirect transmission (which involves an intermediate carrier like an environmental surface or piece of medical equipment).

Portal of Entry
The opening where an infectious disease enters the host's body such as mucus membranes, open wounds, or tubes inserted in body cavities like urinary catheters or feeding tubes.

Susceptible Host
The person who is at risk for developing an infection from the disease. Several factors make a person more susceptible to disease including age (young people and elderly people generally are more at risk), underlying chronic diseases such as diabetes or asthma, conditions that weaken the immune system like HIV, certain types of medications, invasive devices like feeding tubes, and malnutrition.

FIGURE 1.8 Each link in the chain of infection transmission represents a critical stage in the infectious process. Breaking any link in the chain can stop an infection.

Virginia Department of Health. (2011). Chain of infection. Retrieved from http://www.vdh.virginia.gov/epidemiology/surveillance/hai/documents/pdf/Chain%20of%20Infection%20Fact%20Sheet.pdf

reservoir. Fruit bats were found in the roof of a cotton factory in Sudan where three of the first victims of the original Ebola outbreak worked. Similarly, in 1998 sixty miners who worked in an underground mine containing more than 30,000 bats died of Marburg virus infection (a virus very similar to Ebola).

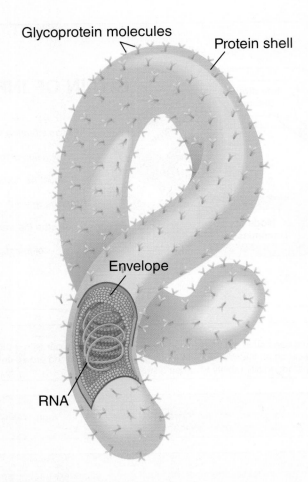

FIGURE 1.9 The structure of an Ebola virion.

During the period between outbreaks of the disease, animal trapping studies were carried out in the rain forests of Central Africa in efforts to determine the natural reservoir. The areas sampled were the same as the locations of outbreaks between 2001 and 2005. Three species of fruit bats were found to be infected with the Ebola virus but did not exhibit symptoms of the disease (Gonzalez, Pourrut, & Leroy, 2007), strongly suggesting that fruit bats are the natural reservoir. Great apes are believed to become infected by feeding on virus-contaminated fruit with spillover to humans from eating or handling contaminated meat.

Bats are the reservoir for several other deadly viruses such as rabies, Marburg virus, and the coronavirus associated with severe acute respiratory syndrome (SARS). Bats may be immune to these viruses for several reasons. Some researchers believe that the bat's body temperature can be elevated through metabolic activity and may mimic the effect of fever in enhancing immune responses. The **genomes** of bats so far studied also carry a large number of virus-related genes as well as a higher number of DNA-repairing enzymes and antiviral proteins that seem to protect them from viral diseases.

Portal of Exit

The Ebola virus is believed to leave bats (assuming that they are the natural reservoir) by saliva or fecal matter. Infection of nonhuman primates is believed to occur

TABLE 1.4	Some Common Zoonotic Infections and Their Reservoirs	
Zoonotic Disease	**Reservoir**	
Viral		
Hantavirus pulmonary syndrome	Deer mice	
Influenza	Birds, swine, bats	
Rabies	Many mammals	
West Nile virus disease	Birds, horses	
Ebola virus disease	Bats	
Bacterial		
Anthrax	Domestic animals	
Plague	Rats	
Lyme disease	Field mice, white-tailed deer	
Rocky Mountain spotted fever	Rodents	
Salmonellosis	Poultry, reptiles	
Other Microbes and Parasites		
Ringworm (fungal)	Domestic mammals	
Malaria	Monkeys	
Toxoplasmosis	Cats, other mammals	
Pork tapeworm	Pigs	

from their eating fruit contaminated with saliva or feces. If fruit bats are not the natural reservoir, then the portal of exit is unknown.

Modes of Transmission

How the Ebola virus is transmitted among bats, the presumptive reservoir, is unknown at the present time. The mode(s) of transmission among the great apes similarly remains unknown but may occur by direct contact as in humans. The transmission of Ebola virus to humans has been shown to occur following eating infected meat from bats or great apes. Human-to-human contact occurs from unprotected direct contact with an infected individual or cadaver. Virus particles have been isolated from all human bodily fluids and secretions. Sexual intercourse is also suspected to be a route for transmission, because virus particles have been isolated from semen up to 5 months after recovery. Interruption of transmission requires the use of personal

© Stephen Mcsweeny/Shutterstock

protective equipment (gowns, goggles, masks, gloves, rubber boots, mask respirator) for all cases of contact with an infected person. Personal protective equipment must be disposed of as infectious medical waste requiring steam autoclaving or incineration.

Portal of Entry

The Ebola virus enters the human body through breaks in the skin, which includes micro-abrasions that are not obvious. It is also able to penetrate mucous membranes throughout the body. The virus enters target cells by a process of fusion with the cell membrane which depends on the binding of a viral glycoprotein with a cholesterol transporter protein identified as Niemann-Pick protein (NPC-1). This is the same transporter that is the cause of Niemann-Pick disease if it is absent. Niemann-Pick disease is a fatal condition in which cholesterol accumulates within cells in the body and cannot be eliminated.

Vulnerable Groups

Healthcare workers caring for those infected as well as family members of an infected person at home represent the highest risk groups. In addition, persons involved with any phase of the preparation of a deceased patient for burial are also at high risk. Persons having direct contact with an infected Ebola patient are also a high-risk group.

Courtesy of National Institute of Allergy and Infectious Diseases (NIAID)/CDC

CRITICAL THINKING QUESTIONS

1. Explain how a recognized illness can be considered an emerging infectious disease.

2. Explain why antibiotic resistance is considered a form of evolution and why this is a concern for international health agencies such as the World Health Organization.

3. How does weather affect the outbreak of an illness?

4. Discuss how a vaccine would interrupt the chain of infection in a disease like Ebola.

5. List at least three reasons why finding the index case of an outbreak is considered important to disease detectives.

6. Why did the 2014 outbreak of Ebola virus disease affect thousands of people while in the past only a few hundred became infected?

7. What group(s) is/are at the highest risk for becoming infected with the Ebola virus and why?

Biology of Ebola

CAROLYN DEHLINGER

"[A] few weeks into this, I've certified the deaths of more patients than in my last two decades. And I'm shocked to the degree to which it has just become part of my daily routine."

—American doctor Joel Selanikio, International Medical Corps,
Lunsar, Sierra Leone

In 1976, a deadly viral outbreak occurred near the Ebola River in what is now the Democratic Republic of Congo (at the time known as Zaire). Named for the river, Ebola outbreaks have peppered African history ever since, with most occurring sporadically and temporarily—until now. According to the **World Health Organization**, as of March 22, 2015, the disease has taken an estimated 10,326 lives worldwide. Ebola was previously known as Ebola hemorrhagic fever. While the name has shortened, the signs and symptoms of this disease have not. The words of Dr. Selanikio remind us of the emotional toll and psychological stress this disease can bring. These aspects can easily be obscured when we discuss the cold facts of a miniscule yet lethal virus. While reading this chapter whose focus is on the biology of the Ebola virus, or more accurately, viruses—to be sure, more than one strain of Ebola plagues the animal kingdom—it is also important to reflect on the human impact.

■ QUESTIONS TO CONSIDER

Some questions to consider as you read this chapter include:

1. What are the symptoms of Ebola and how is it contracted?

2. What is the classification of Ebola, and how is the classification determined?

3. What makes Ebola unique from its viral relatives and what makes it similar?

4. How would you describe the replication cycle of Ebola, and how might details of the replication cycle enable the development of effective antiviral medications to combat Ebola infections?

5. What is the probable reservoir organism of Ebola, and have any other candidates emerged since the time of this writing?

▸▸ Hemorrhagic Viruses

Hemorrhagic viruses are given their name because they cause **viral hemorrhagic fever (VHF)**, a severe syndrome of the body affecting multiple organ systems with multiple symptoms. Ebola is not the only hemorrhagic virus; there are four families of viruses with this designation. The filovirus family, which includes Ebola, is one. Filoviruses are the focus of this chapter. Arenaviruses, bunyaviruses, and flaviviruses are viral families with hemorrhagic members as well. Arenaviruses are transmitted to humans by rodents (**TABLE 2.1**). Bunyaviruses infect rodents, plants, insects, and, to a lesser extent, humans. A well known bunyavirus infecting humans is the hantavirus. Human diseases caused by the flaviviruses include hepatitis C, dengue fever, West Nile virus, and yellow fever. They are often transmitted between mammal hosts through a mosquito or tick vector. **Vectors** are organisms that pass infectious agents among different species. The lifecycle of the infectious agent takes place in part within the vector. A few hemorrhagic viruses and their characteristics are described in **TABLE 2.2** .

The shared symptoms of VHFs typically include:

- Fever
- Muscle aches
- Dizziness
- Fatigue
- Weakness
- Exhaustion

TABLE 2.1	Human Diseases Caused by Arenaviruses
Disease	**Arenavirus**
Argentine hemorrhagic fever	Junin virus
Bolivian hemorrhagic fever	Machupo virus
Brazilian hemorrhagic fever	Sabiá virus
Chapare hemorrhagic fever	Chapare virus
Lassa fever	Lassa virus
Lujo hemorrhagic fever	Lujo virus
Lymphocytic choriomeningitis	Lymphocytic choriomeningitis virus (LCMV)
Venezuelan hemorrhagic fever	Guanarito virus

Source: Data from Easton, A. J. & Pringle, C. R. (2011). Order *Mononegavirales*. In A. M. Q. King, M. J. Adams, E. B. Carstens, & E. J. Lefkowitz, *Virus Taxonomy—Ninth Report of the International Committee on Taxonomy of Viruses* (pp. 653–657). London, UK: Elsevier/Academic Press.

TABLE 2.2	Viral Diseases Causing Hemorrhagic Fevers					
Disease	Causative Agent	Family	Signs & Symptoms	Transmission	Treatment	Prevention
Yellow fever	Yellow fever virus	Flavivirus	Acute phase: Headache, fever, muscle pain Toxic phase: Severe nausea, black vomit, jaundice, hemorrhaging	Bite from a *Stegomyia aegypti* mosquito	No antiviral medications Supportive care	Vaccination Avoiding mosquito bites in endemic areas
Dengue fever	Dengue fever virus	Flavivirus	Sudden high fever, headache, nausea, vomiting	Bite from an infected *Stegomyia aegypti* mosquito	No specific treatment available	Avoiding mosquito bites in endemic areas
Dengue hemorrhagic fever	A different serotype of dengue fever virus	Flavivirus	Decrease in platelets, skin hemorrhaging	Bite from an infected *Stegomyia aegypti* mosquito with another dengue virus	No specific treatment available	Avoiding mosquito bites in endemic areas
Ebola/ Marburg hemorrhagic fevers	Ebola and Marburg viruses	Filovirus	Fever, headache, joint and muscle aches, sore throat, weakness Internal bleeding and hemorrhaging	Bite of an infected fruit bat Blood transfer through cut, abrasion, or infected animal bite	No specific treatment available	Avoiding dead animals and bats in outbreak areas
Lassa fever	Lassa fever virus	Arenavirus	Severe fever, exhaustion, hemorrhagic lesions on throat	Aerosol and direct contact with excreta from infected rodents	Ribavirin	Avoiding dead or infected rodents Maintaining good home sanitary conditions

Infected patients often exhibit signs of bleeding, whether beneath the skin, within internal organs, or from body orifices like the mouth, nostrils, ears, or anus. The term *hemorrhagic* refers to leakage such as the bleeding described. As VHF progresses, more severe cases result in delirium, shock, seizures, or coma. The nervous system can malfunction, contributing to the aforementioned conditions. Common symptoms of Ebola include severe stomach pain, vomiting, and diarrhea.

Taxonomy of Filoviruses

The International Committee on Taxonomy of Viruses is responsible for the classification of known viruses. The taxonomy of viruses is based upon their anatomy, or arrangement of physical structures, including the genetic material they carry. Viruses that possess single-stranded, negative-sense ribonucleic acid (RNA) belong to the order known as *Mononegavirales*. The Greek *monos* refers to the single strand of RNA, while the Latin *negare* indicates the negative polarity of the RNA (see Anatomy of Filoviruses section that follows). The order *Mononegavirales* contains a total of five families:

- *Bornaviridae*
- *Nyamaviridae*
- *Rhabdoviridae*
- *Paramyxoviridae*
- *Filoviridae*

TABLE 2.3 provides details on the family members of order *Mononegavirales*. Ebola belongs to the family of viruses known as the *Filoviridae*, or simply the filoviruses. The family is composed of just three genera: *Cuevavirus*, *Marburgvirus*, and *Ebolavirus*. **FIGURE 2.1** shows a phylogenetic tree of order *Mononegavirales*.

TABLE 2.3	The Family Members of Order *Mononegavirales*		
Family	Number of Genera	Example Viruses	Host Species
Bornaviridae	1	Borna disease virus Avian Bornavirus	Sheep, horses, cattle, birds, rodents, sometimes humans
Filoviridae	3	*Ebolavirus* *Marburgvirus* *Cuevavirus*	Bats, primates, humans
Nyamaviridae	1	Nyamanini virus Midway virus	Ticks, birds

(continues)

TABLE 2.3	The Family Members of Order *Mononegavirales* (Continued)		
Family	**Number of Genera**	**Example Viruses**	**Host Species**
Paramyxoviridae	7	Atlantic salmon paramyxovirus Avian metapneumovirus Fer-de-lance virus Hendra virus Human respiratory syncytial virus Measles virus Mumps virus Newcastle disease virus Sendai virus	Humans, most vertebrates
Rhabdoviridae	11	Rabies virus (RABV) Lettuce necrotic yellows virus (LNYV) Bovine ephemeral fever virus (BEFV) Infectious hematopoietic necrosis virus (IHNV) Potato yellow dwarf virus (PYDV) Vesicular stomatitis Indiana virus (VSIV)	Vertebrates, invertebrates, plants

Source: Data from Easton, A. J. & Pringle, C. R. (2011). Order *Mononegavirales*. In A. M. Q. King, M. J. Adams, E. B. Carstens, & E. J. Lefkowitz, *Virus Taxonomy—Ninth Report of the International Committee on Taxonomy of Viruses* (pp. 653–657). London, UK: Elsevier/Academic Press.

Marburgvirus was the first recognized filovirus, named after Marburg, Germany, where in 1967 an index case occurred in humans (FIGURE 2.2). Laboratory workers handling tissue from green monkeys came down with hemorrhagic fever, resulting in 31 cases and seven deaths.

As described earlier, in 1976 the first known cases of Ebola hemorrhagic fever were recorded. In addition to the case in Zaire, another outbreak occurred concurrently in Sudan. These were the initial strains of Ebola recognized, with later outbreaks revealing still more distinctive species. The strain causing the first ever human outbreak is the same one responsible for the most recent outbreaks in Guinea, Sierra Leone, DRC, and

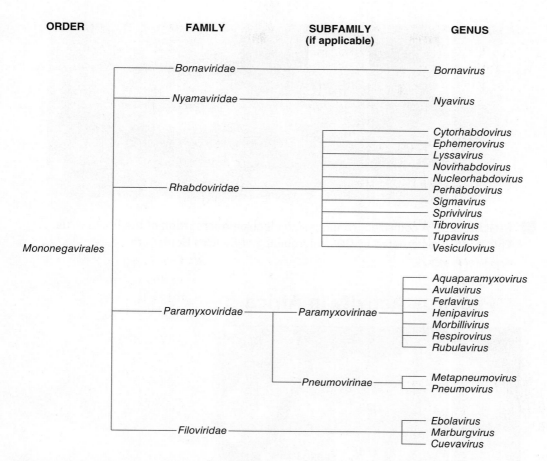

FIGURE 2.1 The phylogenetic tree of order *Mononegavirales* shows the five member families and the genera each contains. Note the location of *Ebolavirus* within the tree.

Source: Data from Easton, A. J. & Pringle, C. R. (2011). Order *Mononegavirales*. In A. M. Q. King, M. J. Adams, E. B. Carstens, & E. J. Lefkowitz, *Virus Taxonomy—Ninth Report of the International Committee on Taxonomy of Viruses* (pp. 653–657). London, UK: Elsevier/Academic Press

Liberia, now considered a widespread epidemic. FIGURE 2.3 shows an image of the Ebola virus.

Cuevavirus is the most recent addition to the filovirus family. Reported in 2010, the single identified species *Lloviu cuevavirus* was discovered when colonies of *Miniopterus schreibersii* (Schreiber's bats) experienced sudden massive deaths in Portugal, France, and Spain. This genus is not known to infect humans and is named for the cave Cueva del Lloviu in Asturias, Spain, where infected bat carcasses were collected for study. In addition to being the most recently discovered filovirus, *Cuevavirus* is also unique with outbreaks limited to Europe. While the other filoviruses, *Marburgvirus* and *Ebolavirus,* have caused infections in Europe and other regions of the world, thus far most of their damage has occurred on the continent of Africa (FIGURE 2.4).

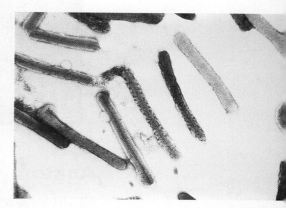

FIGURE 2.2 A transmission electron micrograph showing multiple *Marburgvirus* virions.

Courtesy of R. Regnery; Dr. Erskin L. Palmer/CDC

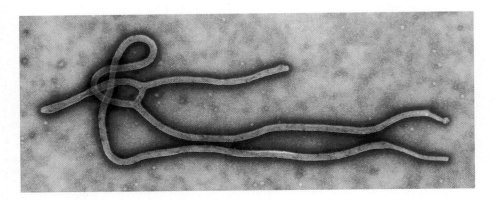

FIGURE 2.3 A colorized transmission electron micrograph of the Ebola virus. This image was created by CDC microbiologist Cynthia Goldsmith.

Courtesy of Cynthia Goldsmith/CDC

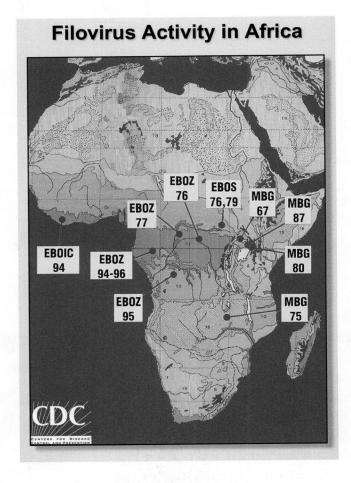

FIGURE 2.4 The distribution of reported cases of *Marburgvirus* (indicated as MBG) and *ebolavirus* (indicated as EBOZ, EBOIC, and EBOS) on the African continent.

Courtesy of CDC

Anatomy of Filoviruses

All viruses share certain anatomical traits. At minimum, viruses possess some form of genetic material enclosed in a protein layer called a **capsid**. Collectively, the capsid and genetic material may be called the **nucleocapsid**. Some viruses also have an additional layer of lipids. These viruses are referred to as enveloped, because the lipid layer is termed an **envelope**. The genetic material of a virus may be deoxyribonucleic acid

(DNA) or RNA in an assortment of forms. DNA viruses exist in either double-stranded or single-stranded genomes. RNA viruses may also be single or double stranded. When a virus contains single-stranded RNA (ssRNA), the genetic material may be deemed positive sense or negative sense. With **negative-sense** RNA, the molecule possesses a base sequence complementary to the viral messenger RNA (mRNA). In **positive-sense** RNA viruses, the viral genome is identical to its mRNA transcript and thus can undergo direct translation into its protein products. Another way to look at the sense of ssRNA is that it constitutes the polarity of the molecule.

Genetically, filoviruses contain negative-sense, single-stranded RNA. Therefore, during gene expression, a filovirus genome must be transcribed into its complementary mRNA by the enzyme RNA replicase before translation can occur. Filovirus nucleocapsids are enveloped.

Filovirus **virions**, or whole viral particles, take on a number of shapes in their appearance. This trait is known as **pleomorphism**. Filoviruses may be circular, u-shaped, short curlicues, or long, sometimes branched, filaments. The longest observed filoviruses reach up to 14,000 nanometers in length and around 80 nanometers in diameter. Each virion consists of a single-stranded RNA molecule, serving as the viral genetic material (or **genome**) surrounded by a lipid envelope. For example, compare the transmission electron micrograph (TEM) images of *Marburgvirus* (shown previously in Figure 2.2) with FIGURES 2.5a and 2.5b. Even when observing the same genus, pleomorphism is apparent.

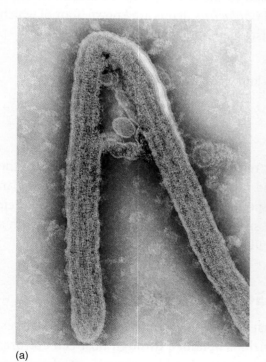

(a)

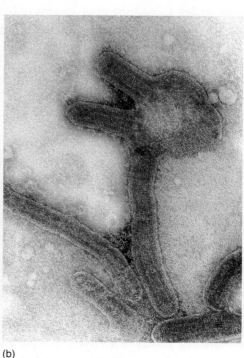

(b)

FIGURE 2.5 (a) This transmission electron micrograph depicts a *Marburgvirus* virion. This image was captured in 1968 by F. A. Murphy, who had grown the virus in host cell culture. (b) Another transmission electron micrograph of *Marburgvirus* taken by F. A. Murphy. Compare this image to Figures 2.2 and 2.5a.

Courtesy of Frederick Murphy/CDC

Transmission of Filoviruses

How does someone contract Ebola or any other filovirus? Direct physical contact between individuals is one way, or, in lieu of direct physical contact, one individual may pick up the virus through contact with the body fluid(s) of an infected person (or animal). There is some experimental evidence that Ebola may be transmitted through small airborne particles, but this has not been directly observed in actual cases. All filovirus outbreaks show the greatest rates of infection occur among patient care-givers, followed by cross-contamination in patient care settings such as hospitals. The latter is known as **nosocomial transmission**. Frequently, these types of transmissions are the result of reusing contaminated needles or syringes. The manner in which filoviruses like Ebola are transmitted makes patient isolation a critical factor in controlling the spread of infection.

▶▶ Characterization of Ebola

The *Ebolavirus* genus has been characterized based on morphological studies using electron microscopy. These observations lend to the development of diagnostic techniques. At least five species or strains of the *Ebolavirus* genus are known, four of which (indicated here with an asterisk, *) cause disease in humans:

- *Bundibugyo ebolavirus* (BEBOV)*
- *Cote d'Ivoire ebolavirus* (CIEBOV), also known as Taï Forest Ebola*
- *Reston ebolavirus* (REBOV)
- *Sudan ebolavirus* (SEBOV)*
- *Zaire ebolavirus* (ZEBOV)*

Thus, the Reston species of Ebola does not result in disease in humans. Studies of the first outbreaks of Ebola in 1976 determined they were caused by the Zaire and Sudan species, respectively, and indeed these two species are named after the locations of their first appearances in humans. Both of these species are highly lethal to humans, with a 90% fatality rate in Zaire cases and 50% fatality rate in Sudan.

Ebola Morphology and Genetics

Morphological studies show all Ebola viruses are distinct from their closest relatives, the Marburg genus of viruses. Marburg virions obtained through host cell cultures were shorter in length than all Ebola virions. During biosynthesis, *Marburgvirus* develops inclusion bodies within host cells of unique morphology as compared to their *Ebolavirus* cousins. While it is difficult to differentiate *Ebolavirus* variants from one another based on inclusion body morphology, only the *Zaire ebolavirus*

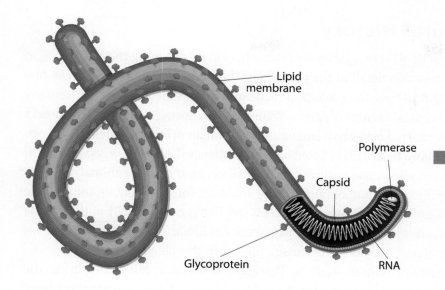

FIGURE 2.6 The generalized anatomy of all Ebola viruses includes a single-stranded RNA genome, one or more RNA replicase enzymes, a lipid envelope, and glycoprotein projections.

© ttsz/iStockphoto

species of Ebola formed inclusion bodies composed of filoviral matrix proteins and nucleoproteins. The association between these proteins continued into the assembly stage of the ZEBOV lifecycle, also unique to this species.

All Ebola species are composed of an ssRNA genome and one or more RNA replicase enzymes surrounded by a lipid envelope studded with glycoproteins (**FIGURE 2.6**). The Ebola nucleocapsid is intertwined with the viral proteins NP, VP35, VP30, and L. Glycoprotein (GP) spikes are embedded on the Ebola envelope, while the space in between the nucleocapsid and envelope are occupied by the viral proteins VP40 and VP24. There are a total of seven genes in the 19-kilobase Ebola genome, listed in 3′ to 5′ order as follows:

3′-leader-NP-VP35-VP40-GP/sGP-VP30-VP24-L-trailer-5′

Genome sequencing studies conducted by Broad Institute and Harvard University researchers in fall 2014 revealed a number of variations in *Ebolavirus* sequences among a sample of Sierra Leone patients. During the first month of the outbreak in Sierra Leone, 78 patients provided blood samples allowing for a total of 99 genome sequences. For some patients, multiple samples were taken over a span of time in order to observe any changes that may occur during the course of disease progression. A total of 395 mutations were discovered among the sequencing data, indicating rapid change in the viral population. Of these, more than 340 are unique to the current widespread epidemic and 50 are exclusive to the West Africa outbreaks.

The sequencing data prove the ongoing widespread epidemic is genetically distinct from the first Ebola outbreaks in 1976, indicative of mutations accumulating over time. The data also suggest there was a single infection—a common ancestor infecting one person—dating back to 1976. Therefore, although today's strain of Ebola is not genetically identical to the 1976 version, it is likely a direct descendant. Genome sequencing has revealed current *Ebolavirus* populations diverged from the original population sometime within the last decade. Furthermore, the researchers were able to determine the virus was spread from Guinea to Sierra Leone by 12 people, all of whom attended the same funeral.

Ebola Natural History

An understanding of Ebola ecology may aid researchers in determining the animal origin of the disease (discussed in detail in the paragraphs that follow). Based on outbreak history, Ebola is thought to have the ecological niche of rainforest ecosystems, particularly in the western and central regions of Africa. Transmissions during the past 17 outbreaks (1996–2007) have occurred between humans and nonhuman primates, antelopes known as duikers, and, possibly, bats. This is based on RNA sequence data of *Zaire ebolavirus* infections in human and serological analysis of mammalian **reservoirs**, or animal of origin, for the ZEBOV-specific immunoglobulin G (IgG) antibody. Interestingly, an infographic published by the **Centers for Disease Control and Prevention (CDC)** outlining the ecology of Ebola displays bats as the primary reservoirs of Ebola (FIGURE 2.7). They are the primary candidates at present, as discussed further in this chapter.

A 2004 University of Kansas study by Peterson, Bauer, and Mills comparing the phylogeny and ecological niche characteristics of the filovirus family proposed relationships among the Ebola species. *Zaire ebolavirus* and *Cote d'Ivoire ebolavirus* appear most closely related, with *Sudan ebolavirus* being most distantly related to all other species, followed by *Reston ebolavirus*.

Ebolavirus Ecology

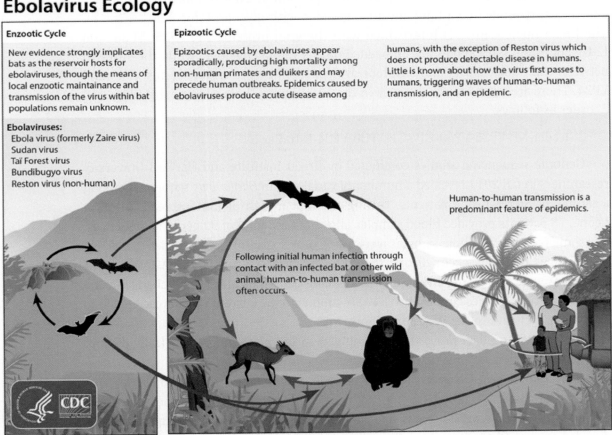

Enzootic Cycle

New evidence strongly implicates bats as the reservoir hosts for ebolaviruses, though the means of local enzootic maintainance and transmission of the virus within bat populations remain unknown.

Ebolaviruses:
Ebola virus (formerly Zaire virus)
Sudan virus
Taï Forest virus
Bundibugyo virus
Reston virus (non-human)

Epizootic Cycle

Epizootics caused by ebolaviruses appear sporadically, producing high mortality among non-human primates and duikers and may precede human outbreaks. Epidemics caused by ebolaviruses produce acute disease among humans, with the exception of Reston virus which does not produce detectable disease in humans. Little is known about how the virus first passes to humans, triggering waves of human-to-human transmission, and an epidemic.

Following initial human infection through contact with an infected bat or other wild animal, human-to-human transmission often occurs.

Human-to-human transmission is a predominant feature of epidemics.

FIGURE 2.7 A recent infographic produced by the CDC shows bats as the reservoir of Ebola viruses.

Courtesy of CDC

Zaire ebolavirus

Zaire ebolavirus was the first Ebola species discovered, and it also represents the source of the most recent widespread epidemic. **FIGURE 2.8** displays a map of West Africa indicating the distribution of cases within the countries of Guinea, Sierra Leone, Liberia, and Côte d'Ivoire. This information was provided by the World Health Organization (WHO) and is current up to October 10, 2014. Epidemiologists utilize maps such as this one to observe trends in the spread of infection. Treatment strategies can also be influenced by these distribution maps. Note the locations of hospitals, laboratories, transit centers for patients, and WHO-designated Ebola Treatment Units (ETUs). The opening image of this chapter depicts the view from inside one ETU looking out through the entrance.

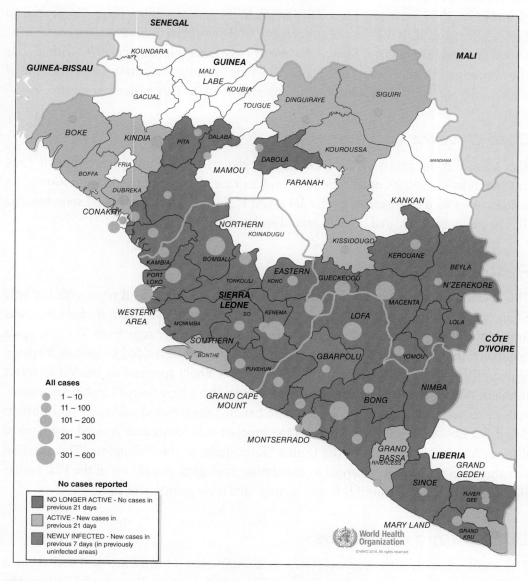

FIGURE 2.8 The distribution of Ebola infections within West Africa as of October 10, 2014, as reported by the World Health Organization.

The Zaire strain of Ebola is historically the most virulent. *Zaire ebolavirus* has resulted in 17 separate outbreaks since the initial outbreak in 1976. Multiple outbreaks have taken place in Zaire (now DRC) and Gabon, with additional outbreaks in South Africa and twice in Russia. The latest outbreak has spread infections into Sierra Leone, Guinea, and Liberia as well. West African countries experience repeated outbreaks that seem to indicate a connection to ecology for this species, whereas infections in South Africa and Russia were caused by exposure to healthcare workers and laboratory researchers, respectively. The widespread nature of the most recent outbreaks has reached epidemic proportions.

Sudan ebolavirus

The second species of Ebola recognized, *Sudan ebolavirus* was first seen in the country of Sudan in 1976, not long after ZEBOV made its appearance in Zaire. Compared to *Zaire ebolavirus*, SEBOV is less fatal, yet more fatal compared to *Bundibugyo ebolavirus*. A total of eight outbreaks of *Sudan ebolavirus* have taken place since 1976, with all but one occurring in either Sudan or Uganda. The lone exception was an isolated infection of a laboratory worker in England, which happened in 1976. This patient survived.

Cote d'Ivoire ebolavirus

Also known as Taï forest virus, *Cote d'Ivoire ebolavirus* was not observed until 1994 in Côte d'Ivoire (Ivory Coast). The lone human case was a scientist who performed an autopsy on a wild chimpanzee in the Taï forest region of Côte d'Ivoire and soon became ill. The patient was treated in Switzerland and survived the infection.

Reston ebolavirus

This species is perhaps the most enigmatic in terms of its origin, as it represents the only Ebola species initially observed outside of the African continent. *Reston ebolavirus* was found in 1989 when captive macaques came down with hemorrhagic fever. The macaques originated from a breeding facility in Luzon, Philippines, but resided in Reston, Virginia, lending this species of Ebola its name. It is the only Ebola species that has yet to infect humans. After seven separate contamination events, there have been 13 instances of exposure in humans. One laboratory worker who handled the infected macaques postmortem developed Ebola antibodies, but none of the exposed individuals ever showed any signs of disease. All cases took place in the United States, Italy, or the Philippines. The question remains whether REBOV emerged in the similar ecological conditions of the Philippines (i.e., tropical rainforests) or if this species migrated from somewhere else.

Bundibugyo ebolavirus

The most recently discovered species is *Bundibugyo ebolavirus*. Isolated from a 2007–2008 outbreak in the Bundibugyo district of western Uganda, BEBOV is the least fatal within the genus with a mortality rate of 34%. Despite this, BEBOV results in twice the rate of bleeding manifestations in patients compared to other strains of Ebola. Only one other outbreak of *Bundibugyo ebolavirus* has taken place. In 2012, 36 individuals became infected with BEBOV in DRC. Although this occurred concurrently with an outbreak of *Sudan ebolavirus* in Uganda, there is no epidemiological evidence linking the two events.

Paleovirology of *Mononegavirales*

During the course of more than 3 billion years of evolution of life on this planet, species have accumulated snippets of viral genes within their genomes. Think of them as footprints implanted into the host cell DNA, left behind by viral invaders of long ago. Observation of these viral remnants offers insight into the ancient relationships between viruses and their hosts. One collaborative study among researchers at the Institute for Advanced Study at the Simons Center for Systems Biology (Princeton, New Jersey); the Fox Chase Cancer Center, Institute for Cancer Research (Philadelphia, Pennsylvania); and the University of California, Irvine determined ssRNA viruses of the order *Mononegavirales* have deposited artifacts of their past presence in 19 vertebrate species (Belyi, Levine, & Skalka, 2010). Within the order, the families *Filoviridae* and *Bornaviridae* are almost exclusively represented in these 19 species at nearly 80 separate genome locations. The earliest integration event is estimated at approximately 40 million years ago. This tells us that members of order *Mononegavirales* have been infecting vertebrate species for at least that long. Based on sequence data, the integrated viral artifacts are linked to genetic coding for matrix proteins, glycoproteins, viral nucleocapsids, and RNA replicase. These are the very genes expressed in the filovirus lifecycle (discussed later). Why would vertebrate genomes hold onto relics of viral RNA? Based on the principles of natural selection, it is a reasonable conclusion that they must provide some biological benefit. Perhaps their presence somehow influences their relative virulence upon an infected host in present time, or even whether a species is resistant or susceptible to infection. If there does prove to be a correlation between the presence of endogenous *Mononegavirales* sequences and virulence, this would give the virus an advantage, too. A species resistant to infection would serve the critical function of acting as the viral reservoir, thus allowing the persistence of the virus.

Research Classification

The CDC (2009) classifies Ebola viruses as Biosafety Level 4 (BSL-4) infectious agents, a class of "dangerous and exotic agents that pose a high individual risk of aerosol-transmitted laboratory infections and life-threatening disease that is frequently fatal, for which there are no vaccines or treatments, or a related agent with unknown risk of transmission" (p. 45). There are a total of four safety levels in ascending order, placing Ebola among infectious agents with the highest amount of risk posed to researchers handling this genus of viruses. Hemorrhagic viruses like Ebola are investigated by the Viral Special Pathogens Branch of the CDC (FIGURE 2.9). Their handling requires the most stringent protocols to prevent contamination.

FIGURE 2.9 In this image from the Centers for Disease Control and Prevention, a microbiologist looks out from a Biosafety Level 4 (BSL-4) laboratory.

Courtesy of James Gathany/CDC

▸▸ The Replication Cycle of Filoviruses

Viruses are **obligate parasites**: They require a host for their survival. However, we refer to them as parasites in a loose sense, because they are not technically alive at all. As a matter of course, they are acellular (not made of cell[s]) and thus violate the cell theory of biology, which states:

1. All organisms are composed of at least one cell.
2. All cells come from preexisting cells.
3. The cell is the smallest form of life.

In order to replicate, viruses must associate with a host cell. Generally, a virus penetrates the host cell with its genetic material, which may be DNA or RNA (either single or double stranded), and takes advantage of gene expression mechanisms present within. A gene in the most basic sense provides information to create a protein; think of it as a recipe. When a gene is expressed, the cell is "cooking" the recipe. Gene expression involves two steps: transcription and translation. Transcription creates a messenger RNA molecule from a DNA, or sometimes RNA, template. Translation turns the mRNA molecule into a protein product. These two steps comprise what is referred to as the central dogma of molecular biology (**FIGURE 2.10**). By hijacking the host cell, it unwittingly creates copies of the virus, resulting in its own death as the newly formed virions burst through in their escape, onward to infect new host cells and repeat the whole process.

Viruses have a number of replicative strategies that depend primarily on the type of genetic information they carry. As viruses rely on host cell machinery to complete their

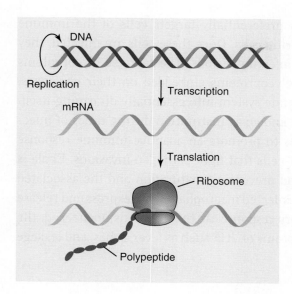

FIGURE 2.10 The central dogma of molecular biology describes the flow of genetic information from DNA to messenger RNA during transcription, then from messenger RNA to a polypeptide (or protein) product during translation.

replication cycles, their mode of replication is dependent upon the presence of or, as we shall see in the case of filoviruses, the lack of necessary enzymes integral to carrying out expression of the viral genome.

Attachment and Entry into Host Cell

Electron microscopy studies of different Ebola species revealed the virus enters host cells via endocytosis. **Endocytosis** is the engulfment of a substance into a cell (FIGURE 2.11). The endocytosis is cell-receptor mediated, which requires the virus to attach to the host cell membrane via receptors on its surface. Specifically, the virus exhibits glycoproteins that enable binding. The host cell membrane begins to invaginate, creating a cavity that eventually pinches off to enclose the virus in a vesicle. The virus has now entered the host cell. From this point, the viral membrane fuses with the vesicle membrane, releasing the viral genome into the host cell's cytoplasm (liquid interior).

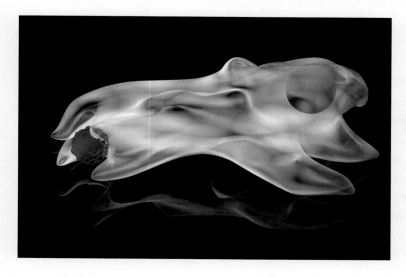

FIGURE 2.11 This model demonstrates the process of endocytosis. The large white cell is surrounding the smaller red substance with its membrane. Ultimately, the red substance will become completely engulfed and pinch off into the white cell.

Research has shown that *Ebolavirus* preferentially targets cells of the immune system, including macrophages and dendritic cells. Once these cells are infected, they release glycoproteins synthesized by the infecting virions. The secreted glycoproteins (sGPs) mimic the behavior of the viral glycoproteins embedded on their envelopes. In this way, *Ebolavirus* confuses the immune system into essentially disarming itself. Dendritic cells normally undergo what is known as maturation during times of infection, during which they release cytokines to promote an adaptive immune response and also trigger the activation of helper T-cells that fight pathogen invasions. Ebola is capable of deregulating dendritic cells and preventing maturation and the associated adaptive immune response. Furthermore, infected macrophages overexpress and release cytokines, which promote an inflammatory response. As this proceeds unchecked, the cytokine storm causes the associated symptoms of VHF such as fever, aches, and leakage of blood into tissues.

Negative-Sense RNA Replication

Filovirus host cells do not normally undergo RNA replication, because their genomes are composed of DNA. The RNA replicase enzyme needed for the virus to complete its replication cycle is lacking. Negative-sense ssRNA viruses such as *Ebolavirus* circumvent this roadblock by carrying one or more copies of the RNA replicase enzyme with them. Upon entering a host cell, the virion piggybacks with the RNA replicase enzyme(s), which then commences making positive-sense strands using the negative-sense strand as a template. These positive-sense strands act as complementary mRNA copies of the viral genome. The genetic information carried within the positive strands can now be used as templates to synthesize more RNA replicase enzymes as well as other viral proteins necessary for reproduction. Indeed, the synthesized positive strands serve as intermediates in the creation of new negative-sense ssRNA offspring (FIGURE 2.12). It is believed the individual genes of the viral genome are transcribed in sequential order by the action of RNA replicase recognizing start and stop regions flanking each gene. FIGURE 2.13 summarizes these steps of the filovirus lifecycle. As noted, this lifecycle is common to negative-sense ssRNA and double-stranded RNA viruses.

Release

Once new progeny are produced in the host cell, they move toward the host cell membrane. In *Zaire ebolavirus,* the matrix protein VP40 appears to play a critical role in interacting with the host cell membrane beneath its surface, allowing viral exit by budding (FIGURES 2.14a and 2.14b). The virions become enveloped in a lipid membrane of their own and are now ready to infect neighboring cells of the host organism. VP40 is experimentally linked to at least two host cell **endosomal sorting complexes required for transport (ESCRT)**. In the broadest sense, ESCRTs are involved in any activity that requires remodeling of the cell membrane.

The tumor susceptibility gene 101 protein is encoded by the human gene *TSG101* and is part of the ESCRT-I complex. This particular ESCRT aids in the regulation

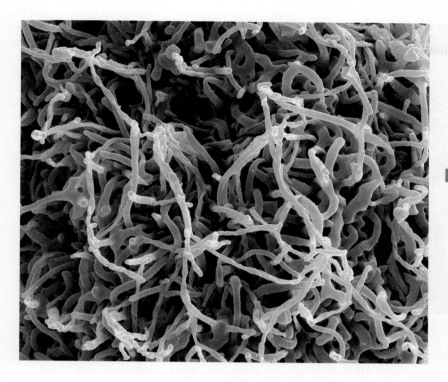

FIGURE 2.12 This scanning electron micrograph (SEM) from the National Institute of Allergy and Infectious Diseases (NIAID) shows a multitude of newly replicated *Ebolavirus* virions within a cultured VERO E6 cell. VERO E6 is a cell line derived from kidney epithelial cells of green monkeys. The image is magnified 50,000×.

Courtesy of National Institute of Allergy and Infectious Diseases (NIAID)/CDC

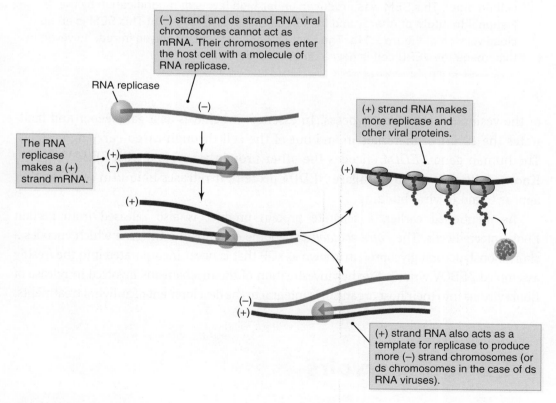

(−) strand and ds strand RNA viral chromosomes cannot act as mRNA. Their chromosomes enter the host cell with a molecule of RNA replicase.

RNA replicase

The RNA replicase makes a (+) strand mRNA.

(+) strand RNA makes more replicase and other viral proteins.

(+) strand RNA also acts as a template for replicase to produce more (−) strand chromosomes (or ds chromosomes in the case of ds RNA viruses).

FIGURE 2.13 The replication cycle of negative-sense, single-stranded RNA viruses requires transcription of positive-sense strands before translation can occur. The positive-sense strands act as mRNA transcript templates, which are translated into gene products.

Note: ds stands for double-stranded.

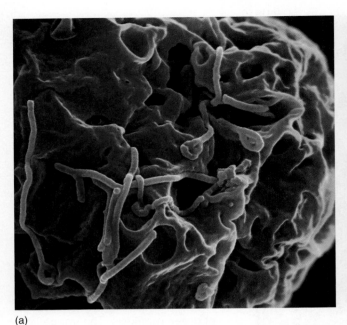
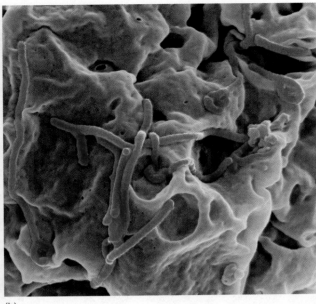

(a)   (b)

FIGURE 2.14 (a) This scanning electron micrograph (SEM) has been digitally colorized to differentiate Ebola virions (in red) budding from the surface of a VERO cell (in blue). This SEM was produced under high levels of magnification by the National Institute of Allergy and Infectious Diseases (NIAID). (b) This SEM is an up-close version of Figure 2.14a. The Ebola virions are still displayed in red; however, in this image the VERO cell is now a taupe coloration.

Courtesy of National Institute of Allergy and Infectious Diseases (NIAID)/CDC

of the vesicular trafficking process. In essence, it behaves as a gatekeeper and facilitates the escort of materials in and out of the cell through cargo-carrying vesicles. The human gene *NEDD4* encodes the other protein associated with ZEBOV's VP40. Known as the E3 ubiquitin ligase NEDD4 protein, it releases ubiquitin proteins that appear to aid in viral budding.

As mentioned earlier, a separate protein product is also released from certain Ebola-infected cells. The *Zaire ebolavirus* genome contains the GP gene which encodes a short, nonstructural glycoprotein known as sGP that is never incorporated into the freshly assembled ZEBOV virions. Further investigation of the mechanisms involved in release of Ebola virions into their host organism is integral to the development of antiviral treatments.

▸▸ Ebola Reservoirs

Ebola is a **zoonotic disease**, meaning it can be spread from animals to humans. One unknown in our understanding of Ebola is the animal origin, or reservoir, of transmission. How exactly the virus first made contact with *Homo sapiens* is still a mystery; however, particulars of *Ebolavirus* outbreaks have revealed distinct clues.

It Is What It Isn't

First, index cases of Ebola are relatively rare compared to other viruses. Furthermore, the index cases of filoviruses tend to be isolated rather than occurring in multiples at once. While some more common viral infections are transmitted through insect vectors, it is not likely for a member of the arthropods to harbor and pass on Ebola. If that were so, we would expect more index cases simply because human contact with insects is so frequent.

In addition, phylogenetic analyses of the filovirus family show a positive correlation to geographic region, meaning each viral strain is strongly associated with a particular location. This observation suggests a degree of stability for the Ebola reservoir. It is likely the reservoir maintains permanent residence or else we would expect there to be a wider geographic distribution of each strain. Building on this logic, if a reservoir is firmly rooted in its habitat, then there is an increased probability of resistance or immunity to the virus and therefore low death rates of the reservoir species. As infected nonhuman primates exhibit high mortality rates, they are generally not considered prime reservoir candidates.

Experimental studies have attempted infecting various species with Ebola. A summary of their results is presented in **TABLE 2.4**. Species that suffer high rates of death are probably not a reservoir, because a reservoir needs to survive in order to continuously introduce the disease to new individuals.

The Leading Hypothesis

One of the primary suspects is the fruit bat. Often, disease reservoirs are species hunted by humans as food sources, and fruit bats are commonly hunted in Africa. Bolstering

TABLE 2.4	Experimental Ebola Infections of Various Organisms	
Organism	**Inoculation**	**Resulting in Infection?**
Plants	*Zaire ebolavirus*	No
Arthropods	*Zaire ebolavirus, Reston ebolavirus*	Very little
Reptiles	*Zaire ebolavirus*	Single infection; low levels of further circulation within the population
Rodents	*Zaire ebolavirus*	Yes
Bats	*Zaire ebolavirus*	Yes; many do not become ill, no deaths
Nonhuman Primates	*Zaire ebolavirus*	Yes; highly lethal

Source: Swanepoel R, Leman PA, Burt FJ, Zachariades NA, Braack LEO, Ksiazek TG, Experimental inoculation of plants and animals with Ebola virus. Emerg Infect Dis. 1996; 2:321–5.

this hypothesis is the recent discovery of Ebola nucleic acid sequences residing in DNA samples derived from at least three fruit bat species. This is indicative of prior contact, because Ebola viruses have the potential to leave behind bits of RNA in its hosts (see the section Paleovirology of *Mononegavirales*, earlier in this chapter). Furthermore, experimental studies have shown bats are capable of surviving Ebola infection.

Typically, a reservoir is identified when an outbreak of disease is observed in a particular wildlife population. Indeed, previous Ebola outbreaks were observed in tandem with outbreaks in other species. Scientists from the Robert Koch Institute and the Wild Chimpanzee Foundation recently monitored wildlife communities surrounding southeastern Guinea, where the outbreak likely originated in humans (FIGURE 2.15). While they have not seen any signs of decline in any local populations of bats or other species, such as chimpanzees and antelopes, they did find a suspicious clue in the village of Meliandou. Just 50 meters from where patient zero, a young boy, lived, a burnt out tree stands (FIGURE 2.16). Villagers told the Robert Koch researchers this tree once contained a population of insectivorous Angolan free-tailed bats (*Mops condylurus*), which local children would play with, as well as capture and barbeque (FIGURE 2.17).

Could the transmission of Ebola to the 2-year-old patient zero have taken place in this tree? Fruit bats are not found in Meliandou, suggesting other potential candidates for transmission are nonexistent. Additionally, out of 169 fruit bats captured in surrounding areas and tested for Ebola by the research team, none carried the virus. Lastly, the first individuals to contract Ebola in Meliandou were children and women, increasing the likelihood that hunting and contact with infected **bushmeat**, usually carried out by men of the village, was not how transmission occurred. All of these indicators combined pointed away from fruit bats. The researchers went on to consider the smaller, insect-eating *Mops condylurus* described by villagers. Previous studies have found Ebola antibodies in this species, which may be evidence of infection. FIGURE 2.18 shows a map of Gabon and DRC, two countries southeast of Guinea, where this discovery was made. Furthermore, sampling the remains of the tree revealed DNA fragments of the Angolan free-tailed bats, confirming villager

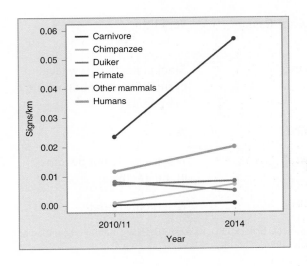

FIGURE 2.15 Densities of various species observed in southeast Guinea, comparing 2010–2011 with 2014.

Robert Koch Institute and the Wild Chimpanzee Foundation. http://embomolmed .embopress.org/content/early/2014/12/29/emmm.201404792.figures-only

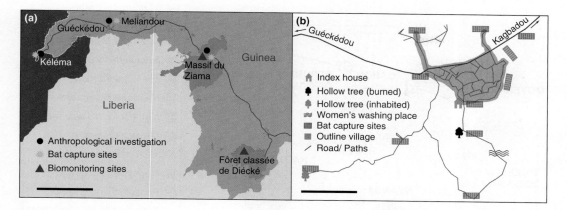

FIGURE 2.17 (a) The village of Meliandou, Guinea. (b–d) The burnt out tree that may have housed Angolan free-tailed bats. (d) A stick near the opening of the tree, which may have been left there by village children who once played close by.

accounts. The first Ebola patient in this epidemic may have played with an infected bat or accidentally ingested the droppings of an infected bat. However, this cannot be definitively proven without further investigation. It is important to note the insectivorous bat hypothesis is still just that.

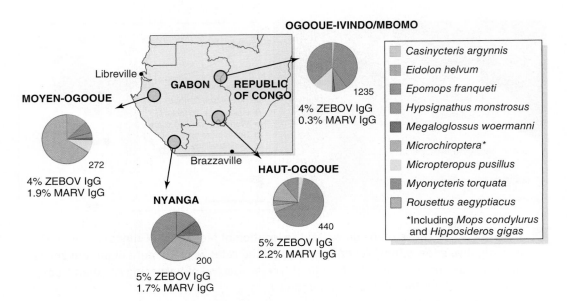

OGOOUE-IVINDO/MBOMO

MOYEN-OGOOUE

Libreville

GABON   REPUBLIC
OF CONGO

1235

4% ZEBOV IgG
0.3% MARV IgG

Casinycteris argynnis
Eidolon helvum
Epomops franqueti
Hypsignathus monstrosus
Megaloglossus woermanni
*Microchiroptera**
Micropteropus pusillus
Myonycteris torquata
Rousettus aegyptiacus

*Including *Mops condylurus*
and *Hipposideros gigas*

272

4% ZEBOV IgG
1.9% MARV IgG

Brazzaville

HAUT-OGOOUE

NYANGA

440

5% ZEBOV IgG
2.2% MARV IgG

200

5% ZEBOV IgG
1.7% MARV IgG

FIGURE 2.18 A map of Democratic Republic of Congo and Gabon, two countries southwest of Guinea, which shows a sampling of bat populations across the region. The pie charts indicate the relative proportions of filovirus antibodies observed in bat specimens, where MARV represents *Marburgvirus* and ZEBOV represents *Zaire ebolavirus*. *Mops condylurus* (Angolan free-tailed bat) is represented in grey within the *Microchiroptera* lineage.

Large serological survey showing cocirculation of Ebola and Marburg viruses in Gabonese bat populations, and a high seroprevalence of both viruses in *Rousettus aegyptiacus*. Xavier Pourrut, Marc Souris, Jonathan S Towner, Pierre E Rollin, Stuart T Nichol, Jean-Paul Gonzalez and Eric Leroy, *BMC Infectious Diseases* 2009, 9:159 doi:10.1186/1471-2334-9-159. © 2009 Pourrut et al.; licensee BioMed Central Ltd.

CRITICAL THINKING QUESTIONS

1. How might the infection mechanisms displayed by Ebola virions affect the progression of symptoms in viral hemorrhagic fever?

2. What is the relationship between the magnitude of the current widespread epidemic and the conditions present in affected countries, and how might these correlations offer suggestions for improvement in response efforts and/or prevention of future outbreaks?

3. How can understanding the evolution of order *Mononegavirales* and its member family *Filoviridae* aid in the development of treatments and/or vaccines? Or understanding Ebola genomics?

CHAPTER 3

Epidemiology

GEORGE EALY

Epidemiology is the study of contagious diseases from the perspective of cause, distribution, and methods for control. The term is derived from the Greek *epi* ("upon") and *demos* ("the people"); in this case, a disease that has fallen upon the people. This chapter discusses the basic concepts of epidemiology in relation to the Ebola virus outbreak of 2013–2015. Emphasis is directed toward the methodologies used by epidemiologists. These include the determination that an outbreak has actually occurred, the development of a standard case definition, the types of epidemiological studies that are used, and methods applied for control. Whenever appropriate, historical background is also introduced.

■ QUESTIONS TO CONSIDER

Some questions to consider as you read this chapter include:

1. Point out the differences and the similarities between incidence and prevalence of a disease.

2. Why is a case definition considered the cornerstone for all epidemiological investigations?

3. Explain the role of person, place, and time in a descriptive epidemiological study and how these relate to the development of an epidemic curve.

4. Why are Koch's postulates not applicable to all microbiological agents?

5. Compare and contrast the Bradford Hill criteria for establishing causality against Koch's and Rivers' postulates.

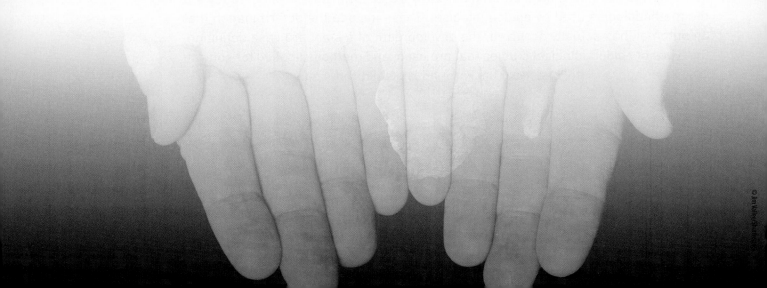

⏵⏵ Establishing an Outbreak

One of the first steps that must be done in investigating the occurrence of an **outbreak** (see TABLE 3.1) is to verify that a suspected outbreak is, in fact, a real outbreak. Although the concept of a disease outbreak may seem self-explanatory, it has a precise meaning in epidemiology. By definition, an outbreak refers to the *occurrence of a disease at an observed rate that is greater than that which is expected* in a group of people. It can also refer to the sudden appearance of a disease in an area that has never experienced the disease before. The term **epidemic** refers to a disease that spreads rapidly and infects many people. An outbreak implies localization, whereas epidemic implies broad spread. Epidemiologists refer to a group of individuals as a population. Two concepts—**incidence** and **prevalence**—are helpful in determining whether or not an outbreak has occurred. Both measure the extent of a disease

TABLE 3.1	Steps in the Investigation of an Infectious Disease Outbreak
Procedure	**Relevant Questions and Activities**
Define the problem	Verify that an outbreak has occurred; is this a group of related cases that are part of an outbreak or a single sporadic case?
Appraise existing data	Case identification: Track down all cases implicated in the outbreak.
	Clinical observations: Record the pattern of symptoms and collect specimens.
	Tabulations and spot maps:
	• Plot the epidemic curve
	• Calculate the incubation period
	• Calculate attack rates
	• Map the cases (helpful for environmental studies)
Formulate a hypothesis	Based on data review, what caused the outbreak?
Confirm the hypothesis	Identify additional cases; conduct laboratory assays to verify causal agent.
Draw conclusions and formulate practical applications	What can be done to prevent similar outbreaks in the future?

Source: Adapted from Friis, R. H., & Sellers, T. A. (2009). *Epidemiology for public health practice* (4th ed.). Sudbury, MA: Jones and Bartlett Publishers; 456–457.

within a population in similar but very different ways. Incidence reflects the change in a population from a nondisease to a disease state and represents the *number of newly diagnosed patients* in a population at risk for the disease. Incidence measures the speed with which a disease appears within a group. As an example, the incidence of Ebola virus infections in the African countries of Guinea and Sierra Leone was almost 2,000 cases per week in the month of November 2014. In contrast, prevalence is a *fixed* or *static* number that represents the number of *both newly diagnosed patients and those who are uninfected* in a population at any given time. Therefore, it is the proportion of a population that has a disease.

A way of understanding both concepts is to imagine a bathtub in which incidence is the rate of water flowing into the tub and prevalence is the proportion of water in the tub at any given moment. Water going down the tub's drain represents patients that have either recovered from the illness or have died. In practical terms, incidence is used more often in describing a disease than prevalence, because incidence reflects the rate at which an illness is spreading within a population.

Another way of expressing the rate at which a disease is spreading within a population is the **reproduction number**, also known as the **rate of infectivity**, both symbolized as R_0. The reproduction number represents the number of secondary infections that occur after an infectious disease is introduced to a susceptible population. The rate of transmission is directly proportional to the value of the reproduction number. The higher the reproduction number is, the greater the possibility of spread. A reproduction rate of 10 means that 10 secondary infections follow one person becoming infected. In the Ebola epidemic of 2014–2015, the reproduction number in the most affected countries was greater than 2, meaning that for each person infected two more individuals became infected. To contain the Ebola epidemic, both the Centers for Disease Control and Prevention (CDC) and the World Health Organization (WHO) aimed for a reproduction number of less than 1, a low enough transmission rate that would bring the epidemic to an end. The frequency with which diseases occur within a population is also used in determining the existence of an outbreak. In any given population, certain diseases are always present at very low frequencies. For example, throughout the calendar year, some members of a population will always have the common cold. These illnesses are considered **endemic** in a population. A disease outbreak may occur against the backdrop of an endemic illness; however, for some rare diseases such as the Ebola virus or Lassa fever the endemic rate is zero. In other words, no cases of these diseases should occur, and any reported case might be considered an outbreak.

Unfortunately, determining disease cases and their numbers can be difficult in economically poor countries with minimally developed healthcare systems. A disease such as Ebola may occur in an isolated village (**FIGURE 3.1**), and significant time may pass before healthcare workers are aware of its occurrence. Additionally, record keeping may also be minimal or nonexistent (**FIGURE 3.2**). Compounding these problems is the fact that many common viral illnesses share signs and symptoms with Ebola, making an accurate diagnosis difficult to determine.

■ **FIGURE 3.1** Photograph of a typical village in Guinea.

Courtesy of Sally Ezra/CDC

Case Definition

For the reasons previously discussed, investigations of a potential disease outbreak are conducted using the concept of a **case definition**. The case definition is one of the most important aspects of any disease investigation. It is a uniform set of conditions that is used to identify an illness for surveillance by public health officials. Its purpose is to produce a standard by which healthcare providers can determine whether or not a person has the disease under investigation. This need arises due to natural variations in how patients describe their symptoms and how those symptoms might be interpreted. A case definition completely describes the disease under study in terms of its clinical signs and symptoms. The **signs** of a disease are *objective findings* such as body temperature, rate of respiration, pulse, blood pressure, or weight. **Symptoms** of a disease are *subjective reports* by a patient that describe their perception of an illness. Statements such as, "I have a headache," or, "I feel feverish," are examples of symptoms. Unfortunately, many illnesses share the same group of signs and symptoms and are not diagnostic for a particular disease. This can make the determination of a diagnosis especially difficult if the disease has a long incubation period. The Ebola virus infection is an example of this problem. It causes a group of symptoms that are shared by many other viral diseases, and patients may not be aware that they are ill for up to 21 days after their first exposure.

■ **FIGURE 3.2** A rural hospital in Guinea lacking modern patient care equipment.

Courtesy of Dr. Heidi Soeters/CDC

TABLE 3.2	Case Definition for Ebola Virus Disease	
	Signs and Symptoms
Person Under Investigation	• Elevated body temperature or subjective complaint of fever • Headache • Muscle aches (myalgia) • Vomiting, diarrhea • Abdominal pain • Unexplained hemorrhages **And:** • A risk factor within the past 21 days before symptom onset*
Confirmed Case	• Laboratory confirmation of infection

*For further information on risk factors consult the CDC website for Ebola Outbreak (http://www.cdc.gov/vhf/ebola/healthcare-us/evaluating-patients/case-definition.html)

The clinical signs and symptoms for the Ebola virus disease were developed by the CDC and include both clinical findings and epidemiological risk factors. The risk factors encompass individuals who are under investigation and those who are actual, confirmed cases. A **person under investigation (PUI)** is defined as someone who has a fever greater than 101.5°F (38.6°C) and additional symptoms of severe headache, muscle pain, vomiting, diarrhea, abdominal pain, unexplained hemorrhage, and a group of associated risk factors that occurred 21 days before the onset of symptoms. The complete case definition for the Ebola virus disease is shown in **TABLE 3.2**.

Epidemiological risk factors for the Ebola virus disease are more complex than the relatively straightforward description of clinical findings and are based on degree and manner of contact with a symptomatic patient. *High-level risk factors* are direct contact with a symptomatic patient in the following situations: exposure to body fluids without protective equipment, a needle-stick injury, contact with the body of a deceased patient, and providing patient care while living in the patient's household. *Moderate-level risk factors* are contacts *without* the use of personal protective equipment (PPE) with a symptomatic patient. *Low-level risk factors* are visiting a country with widespread transmission for longer than 21 days and/or contact with an Ebola patient before the development of symptoms that does not involve exposure to body fluids.

A confirmed case requires laboratory-confirmed diagnostic evidence such as a positive enzyme-linked immunosorbent assay (ELISA) test. The number of confirmed and suspected cases for the 2014–2015 Ebola outbreak as of early April 2015 was reported by WHO as 25,831 in all three affected countries. The number of fatalities in all three affected countries for the same time period was 10,326, yielding a mortality rate of almost 40%.

The Centers for Disease Control and Prevention and the World Health Organization

The CDC and WHO figure so prominently in this chapter that some background information is needed. The CDC (**FIGURE 3.3**) was organized on July 1, 1946, in Atlanta, Georgia, as an insignificant branch of the Public Health Service, with an original mission of keeping the southeastern region of the United States free from malaria. During the 1950s, the CDC played a key role in the nationwide epidemic of poliomyelitis by advocating for the immunization of school-age children with polio vaccine. Disease surveillance and epidemiology became cornerstones for the new agency. The CDC played a vital role in one of the greatest triumphs of modern medicine: the global eradication of smallpox in 1977. It was also responsible in the late 1970s and early 1980s for identifying the bacterial cause of two diseases that at first seemed mysterious and created a nationwide panic: Legionnaire's disease (*Legionella pneumophila*) and toxic shock syndrome (*Staphylococcus aureus*). The CDC was also the first to publish the initial description of HIV (human immunodeficiency virus) infection in 1981 and has continued to publish numerous leading updates on the illness. Each week the CDC also publishes the *Morbidity and Mortality Weekly Report* (*MMWR*), which contains a wealth of information on reportable diseases and outbreaks of illnesses within the United States. The CDC website (www.cdc.gov) provides an alternative source of information for any infectious disease. Since its inception, the CDC has grown to include worldwide assistance in the investigation of disease outbreaks such as the current outbreak of Ebola virus.

WHO is an agency under the auspices of the United Nations. It was created in 1948 and is headquartered in Geneva, Switzerland. WHO acts as a global epidemiological repository and monitors the status of national health delivery systems as well as global health situations. World Health Day, which occurs annually on April 7, focuses attention on major health issues such as vector-borne diseases or antibiotic resistance. WHO has targeted HIV/AIDS (acquired immune deficiency syndrome), malaria, and tuberculosis as diseases for significant reduction. In particular, WHO has as its objective the eradication of poliomyelitis through vaccination, similar to smallpox eradication. Despite these efforts, it recently declared in May 2014 that the spread of polio is a worldwide health emergency, citing outbreaks of the disease in Asia, Africa, and the Middle East as disturbingly significant. More recently, in August 2014 WHO declared the outbreak of Ebola virus disease in West Africa a public health emergency.

FIGURE 3.3 The Centers for Disease Control and Prevention headquarters in Atlanta, Georgia.

Courtesy of James Gathany/CDC

▸▸ Types of Epidemiological Studies

Epidemiology uses three basic approaches to investigate a disease: descriptive, analytical, and experimental. Epidemiologists used all three of these methods in their investigation of the Ebola outbreak. Each of the methods is examined in this section in relation to the Ebola outbreak and epidemic of 2014–2015.

Descriptive Epidemiology

Descriptive epidemiology involves the collection of as much data as possible about the outbreak. The objective of this approach is to identify the source of an outbreak by searching for patterns that are revealed from the data. An examination is conducted in terms of *person, place,* and *time*. The first step in examining the Ebola outbreak was to determine which members of each country were infected and then to look for commonalities among associated variables such as the age of those infected, gender, and ethnic backgrounds. In the case of Ebola, an extra category was considered that included whether or not an infected person was a healthcare provider. Normally, age and gender are considered the most important variables, because they are often related to exposure, and, consequently, to the risk of acquiring an infection. However, in the case of the Ebola outbreak, occupational hazards were considered to be of equal significance.

The analysis of an illness by place can offer insight into the geographical extent of a disease. This information may indicate a clustering of cases, which can provide clues to determining the origin of an infectious disease. Place analysis is a relatively straightforward method for uncovering the existence of any geographical patterns that may be associated with where infected persons live, work, or may have been exposed. The geographical distribution for the Ebola outbreak is shown in FIGURE 3.4 . This distribution established that the outbreak was localized at first but had spread within several months to four countries with contiguous borders. The best explanation for this occurrence is movement of infected but asymptomatic people from one country to another. This led to strict limitation of travel for people living within the affected countries in efforts to contain the spread of the infection.

Characterizing an illness by time is traditionally done by creating a graph of the number of cases beginning with the date of onset and continuing for an interval of time. This generates an **epidemic curve**. An example of an epidemic curve for the Ebola outbreak of 2014 is shown in FIGURE 3.5 . This graph represents the number of cases and deaths between March 22 and August 31. March 22 was chosen as the starting point for the curve, because it coincides with the date WHO was notified that an outbreak of Ebola virus had occurred in Guinea. Interpretation of an epidemic curve provides insight into whether an outbreak occurs from a **point source** or whether an outbreak has progressed into a **propagated outbreak**. If an outbreak occurs from a point source, the number of cases in an epidemic curve rises to a single peak within a generally short

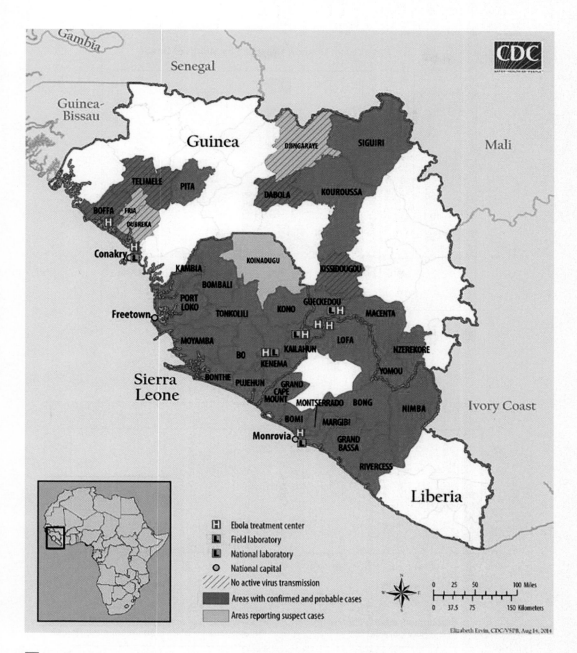

FIGURE 3.4 Map of the distribution of diagnosed Ebola patients in Africa during the 2014–2015 epidemic.

Courtesy of Elizabeth Ervin/CDC

time and then declines. This finding suggests that all of those affected were exposed to a common agent. However, a gradual rise in the number of cases over time suggests that the outbreak is being transmitted by person-to-person contact and is being propagated as an epidemic. **FIGURE 3.6** shows some examples of epidemic curves that give a visual summary of data and can give clues about the nature of an outbreak.

Time can also be analyzed in terms of the **incubation period**, which is the interval between exposure and the onset of symptoms (see **FIGURE 3.7**). Three of the four countries involved in the outbreak (Guinea, Liberia, and Sierra Leone) were examined in this

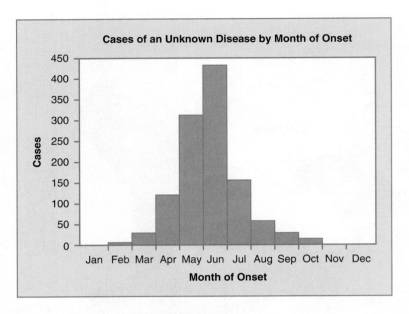

FIGURE 3.5 An example of an epidemic curve.

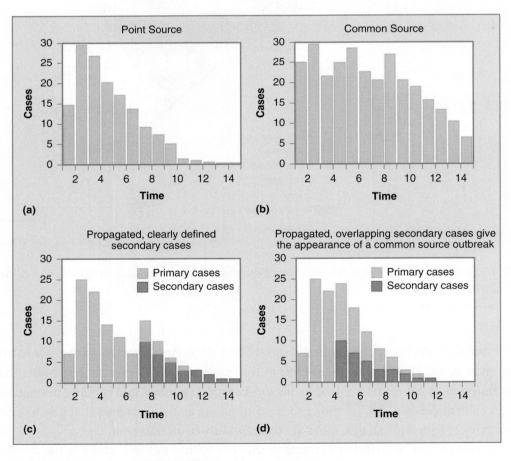

FIGURE 3.6 Examples of epidemic curves illustrating that the shape of the curve can give clues about the mode of transmission and incubation period.

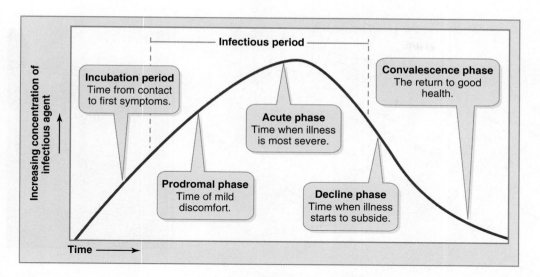

FIGURE 3.7 Most infectious diseases go through a five-phase process: incubation period, prodromal phase, acute phase, decline phase, and convalescence phase. The length of each phase would be dependent on the pathogen and host response.

respect. The average time of incubation was 11.4 days for all three countries; however, the actual time of incubation ranges from a minimum of 2 days up to 21 days. If someone does not develop symptoms after 21 days, they are considered not to be infected with the Ebola virus. However, the validity of the 21-day cutoff period has been challenged by at least one researcher who has presented data indicating that a 31-day period of quarantine is safer for protecting the public.

The incubation period forms the basis for conducting a contact history. In a **contact history** all persons with whom an infected person has had contact are located and interviewed and then followed for the development of symptoms associated with Ebola. The length of time that a contact is followed is 21 days, based on the incubation period, which indicates that a person free of any symptoms after this time is not infected.

Historical Context

Interestingly, the principles used in descriptive epidemiology to determine the parameters of the current Ebola epidemic follow many of the same methods that were developed in 1854 by John Snow in his investigation of a cholera epidemic in London. Snow (1813–1858) was a British physician, who is considered the "father of epidemiology" for his pioneering work in the study of disease transmission. During the mid-19th century, London experienced several epidemics of cholera, which were collectively responsible for more than 20,000 fatalities. Snow was convinced that cholera was spread by water that was contaminated with the fecal matter of cholera victims. In 1849 he published his theory in a treatise, *On the Mode of Transmission of Cholera*. He lived in an era before the germ theory of disease when most doctors believed that diseases were caused by inhaling miasma or "bad air." His concepts were considered radical and rejected by nearly all members of the British medical community.

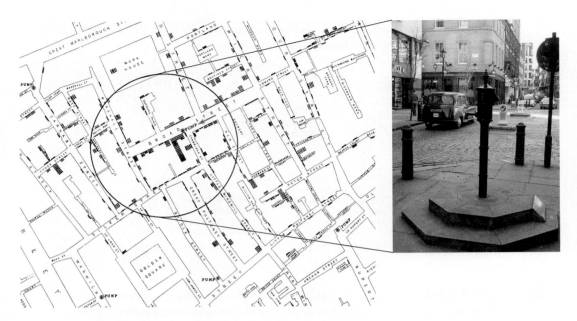

■ **FIGURE 3.8** In London, England, 1854, Dr. John Snow helped interrupt a cholera epidemic by having the handle removed from the pump, located here on Broad Street.

Between August 31 and September 4, 1854, hundreds of people in the Soho neighborhood of London died of cholera, prompting it to be called the "Great Epidemic." Dr. Snow lived near the area and investigated the epidemic, hoping to prove his theory on the transmission of cholera. With the help of a local clergyman, the Reverend Henry Whitehead (1825–1896), Dr. Snow interviewed dozens of cholera victims, which provided the groundwork necessary to describe the characteristics of the illness—the essence of descriptive epidemiology. He constructed a map (FIGURE 3.8) relating the number of fatalities to their proximity to a common water pump located on Broad Street. The Broad Street pump exists today and is memorialized by a plaque describing its role in determining the cause of the great cholera epidemic (see Figure 3.8, inset). Snow combined descriptive and analytical epidemiology to show that the water drawn from the pump was the source of the epidemic. Much as today, he instituted measures of control by persuading the water department to remove the pump's handle. This action dramatically ended the occurrence of new cases within a few days.

Analytical Epidemiology

The goal of **analytical epidemiology** is to determine the parameters of *how* a disease is transmitted from interpreting data acquired during descriptive epidemiological investigations. Data are analyzed to determine the **mode(s) of transmission**, the period of incubation after initial exposure, and the rate and extent that a disease is spreading within a population. In the case of the 2014 Ebola outbreak, several vexing questions exist: How is the disease transmitted, how long is the incubation period, will the disease spread globally, and, if so, how fast?

Experimental Epidemiology

Experimental epidemiology uses mathematical models to analyze interactions of infectious agents with members of a population and determine how rapidly a disease is spreading and how well control measures are working. Several models are commonly used and are designated by the acronyms SI, SIR, SEIS, and SEIR. Because the mathematics used in experimental modeling is beyond the scope of this case example, only the theory behind each model is discussed.

The **SI model** is used to analyze infections where the infected members of a population do not acquire immunity and can be repeatedly infected. The members who are susceptible to becoming infected are designated as "S," and those who are infected are designated as "I." The SI model applies to diseases such as the common cold that are caused by rhinoviruses and coronaviruses.

The **SIR model** is a compartmentalized model where members of a population are divided into three compartments: (S) susceptible members, (I) those infected, and (R) those who have recovered or have been removed from the population by death. This model applies to infections that result in permanent immunity to reinfection following recovery. The SIR model is used to determine the number of people infected with a contagious disease in a population over time. It is used most often with viral infections such as measles.

The **SIRS model** is similar to the SI and SIR models, with the difference being that it measures the rate at which people become reinfected. This model is very useful for sexually transmitted diseases as well as bacterial and worm infections.

The **SEIR model** is used for diseases that are characterized by an incubation period between exposure and the development of signs and symptoms. The "E" represents the incubation period, or the exposed state. Ebola virus falls into the category of the SEIR model.

▸▸ Establishing Cause

Although John Snow could not identify the cause of the cholera epidemic, he did observe living microorganisms in the pump's water, which he believed were the cause. Another 30 years passed before Robert Koch (1843–1910) (**FIGURE 3.9**) would prove that bacteria were the cause of infectious diseases such as anthrax, tuberculosis, and cholera. His findings ushered in the modern era of microbiology. To establish a causal relationship, Koch developed what are called **Koch's postulates**. His postulates remain a cornerstone for epidemiologists in establishing a causal relationship between bacteria and illnesses.

Koch's postulates consist of four steps: *association, isolation, inoculation,* and *re-isolation,* as shown in **TABLE 3.3**. The first step requires that

FIGURE 3.9 Robert Koch developed what became known as Koch's postulates that were used to relate a single microorganism to a single disease.

Courtesy of National Library of Medicine

TABLE 3.3	Koch's Postulates
1. Association	A suspected pathogen must always be associated with the same disease.
2. Isolation	The suspected pathogen must be isolated from an infected animal then grown as a pure culture and identified.
3. Inoculation	A healthy animal is inoculated with a sample from the pure culture and observed for signs of disease.
4. Re-isolation	The suspected pathogen must be re-isolated from the infected animal in step 3 then compared to the organism from step 2 to determine if the organisms are the same.

a suspected pathogen must be consistently associated with a specific illness. This means that the same microorganism must be present in the blood or tissues of all organisms experiencing the disease but absent in those that are healthy. In the second step, the microbe must be isolated from a diseased organism and then grown as a pure culture, meaning that only one type of microorganism is growing. The third step requires inoculation of a healthy animal with the microbe in pure culture. The inoculated animal must then become ill with the same disease. The fourth step requires isolating the same organism from the animal in step three. The fulfillment of all four steps establishes a causal relationship between the microbe and the disease.

Several limitations of Koch's postulates became evident during his lifetime. For example, not all people infected with a microorganism will exhibit signs or symptoms of an illness. This was the case with cholera, the subject of John Snow's investigation. Koch was also aware that some people infected with cholera did not experience symptoms. Another problem with Koch's postulates involves the fact that some microorganisms, such as viruses, are not considered to be alive and cannot be grown in culture. Finally, many pathogens are too small to be seen with a light microscope, which was the most advanced type of magnifying instrument available during Koch's lifetime.

For these reasons, alternative steps have been proposed by several epidemiologists to show causality in situations for which Koch's postulates do not apply. Thomas Rivers, considered the "father of modern virology," developed a set of criteria in 1937 that established viral causes for an illness. **Rivers' postulates** consist of six steps that must be satisfied to establish a viral cause, as shown in TABLE 3.4 . A virus that is filterable must be isolated from an infected host and cultivated. The cultivated virus must cause the same illness in a susceptible animal and also cause an immune response in the form of antibodies.

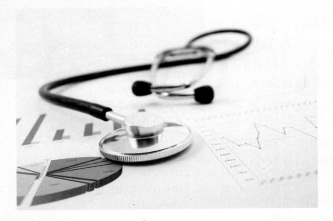

TABLE 3.4	Rivers' Postulates
1.	Every diseased host must contain a virus that can be isolated.
2.	The isolated virus must be able to replicate in cell cultures.
3.	The cultured virus must be able to pass through a filter designed to retain all other microorganisms.
4.	The filtered virus must be able to cause the same disease when inoculated into a healthy host.
5.	The same virus must be isolated from the infected host.
6.	The infected host must develop antibodies to the virus.

However, limitations also exist with Rivers' postulates, because some viruses cannot be cultivated in a laboratory. With advancements in molecular biotechnology, the detection of viral nucleic acid sequences has become the mainstay for establishing the identity of a viral pathogen. Fredericks and Relman in 1996 reformulated both Koch's and Rivers' postulates to include viral nucleic acid sequences in establishing an epidemiological cause for a disease. Their proposal requires the detection of a viral nucleic acid in a diseased host that is paralleled by a decline in its presence with recovery. Additionally, viral sequences should be identified if possible in affected organs. The complete list of requirements is shown in TABLE 3.5 .

In addition to these postulates, epidemiologists use another set of criteria proposed by the British physician, Bradford Hill, to establish cause and effect. Known as

TABLE 3.5	Fredericks and Relman's Postulates
1.	A nucleic acid sequence (either DNA or RNA) must be present in hosts that exhibit signs of a disease.
2.	Disease-free hosts should have no nucleic acid sequences of the putative virus.
3.	Infected hosts recovering from the disease should have either a reduced number or absence of nucleic acid sequences of the putative virus.
4.	A causal relationship is suggested if the nucleic acid sequence occurs in the host before the onset of disease.
5.	Organs or tissues that are affected should demonstrate the presence of nucleic acid sequences by biotechnology methods.
6.	The biotechnology methods used to identify nucleic acid sequences should be reproducible.

Hill's criteria of causation, these requirements represent the minimum conditions that are needed to establish a causal relationship between two items. The essential criterion is based on a temporal relationship in which a proposed cause must always precede an outcome. Thus, if Ebola virus is the cause of a hemorrhagic fever, its presence must always exist before the manifestation of symptoms. Although this seems obvious, it sets aside arguments of cause based on functionality. Furthermore, the Bradford Hill criteria require that a cause of an illness must be supported by statistically valid results; this requirement serves as a keystone for modern epidemiological studies. The complete set of criteria is listed in TABLE 3.6 .

TABLE 3.6	Bradford Hill Criteria for Causation
1. Strength of Association	A small number of findings *does not* eliminate a microbial causation, although the larger the number of positive identifications correlates directly with the likelihood that a disease is of microbial origin.
2. Consistency of Findings	The probability of an infectious origin directly correlates with the consistency of findings by different investigators in different locations with different samples.
3. Specificity	The probability that a microbe causes a disease is enhanced by an illness among a specific population when all other possible causes have been ruled out.
4. Temporality	A disease must occur after its cause.
5. Biological Gradient	An illness must parallel the degree of exposure to a microorganism; the larger the number of people exposed to a microorganism should result in an increase in the incidence of a disease. Also called the dose–response relationship.
6. Plausibility	The development of an infectious disease should have a plausible mechanism for its occurrence.
7. Coherence	A microbial etiology of a disease is strengthened by a coherent correlation between epidemiological and laboratory findings.
8. Experimental	A causal relationship is strengthened by supportive experimental results.
9. Analogy	The results of similar epidemiological studies may be used in considering a causal relationship between a microorganism and a disease.

▶▶ Methods of Transmission

Simply knowing the cause of an illness will not stop its spread. This is why epidemiologists categorize diseases based on how they are transmitted. Two methods of transmission are recognized: direct and indirect contact. Direct contact includes personal contact with an infected or diseased person, which involves touching, kissing, or the exchange of bodily fluids including sexual activity and blood transfusions (see TABLE 3.7). It also includes exposure to airborne particles or aerosols; respiratory droplets from coughing, sneezing, or even talking; direct contact with animals; spread from an infected mother to her fetus (known as vertical transmission); and inanimate objects that are contaminated with a virus and are capable of direct contact by accidental penetration of the skin, such as with an intradermal or hypodermic needle.

Indirect contact involves what epidemiologists call vehicle transmission. This occurs from exposure to substances acting as a vehicle for transmission of a disease such as water, contaminated food, or air currents. Vehicle transmission also involves insects that land on substances and transmit infectious agents by their wings or leg parts. Insects are also a major form of indirect transmission by bites.

Although the CDC has repeatedly asserted that transmission of the Ebola virus only occurs from *direct contact with body fluids* and cannot be spread by airborne routes, some epidemiologists have suggested that the virus may also be transmitted by aerosolized droplets, especially those in the micrometer size range. Early studies of airborne transmission were unable to detect very small particles. When a person sneezes or coughs, particles of varying sizes are released into the air. Studies conducted during the 1940s and 1950s could not detect minute particles that are released

TABLE 3.7	Means of Direct Transmission of Ebola virus
Blood	
Breast milk	
Feces	
Saliva	
Semen	
Vaginal secretions	
Vomitus	

Contaminated inanimate objects:
- Hypodermic needles
- Intradermal needles

close to a person's face during sneezing. For this reason, only large particles were assumed to be infectious. This assumption is reflected by the CDC's recommendation that facemasks (which block large particles) are protective against airborne transmission. However, more recent studies in 2011 indicated that nonhuman primates can be infected with Ebola virus by aerosol particles in the micrometer range. These results were interpreted by some to indicate that respirators (which block small particles) should be worn by healthcare workers when caring for Ebola patients.

A laboratory accident at the CDC in Atlanta in December 2014 exposed several workers *who were not wearing any type of protective face barrier* to the Ebola virus. A sample of live Ebola virus was accidentally transferred from a high-safety laboratory to one that was less secure in which technicians were wearing gloves and gowns but no facemasks. The workers were monitored for 21 days after the accident for the development of Ebola virus disease symptoms. None of the workers became infected with the virus after a 21-day monitoring period.

Length of Quarantine: Is 21 Days Enough?

While the CDC has not revealed the basis for its 21-day quarantine, it is presumably based on analytical studies of earlier Ebola outbreaks that occurred in 1976 (in Zaire, now the Democratic Republic of Congo) and 2000 (in Uganda). However, analytical studies of the incubation period have established that there is a significant **margin of error** regarding the incubation period for Ebola. Charles Haas, a microbiologist at Drexel University, published a study in the *New England Journal of Medicine* that reviewed patient data on all previous outbreaks of Ebola and concluded that 21 days is not sufficient to declare that all people who are asymptomatic after that time are actually free of the virus. His study reported that releasing people from quarantine after 21 days results in a margin of error of up to 12%, which means that as many as 12% of those released may still develop symptoms. For this reason, some epidemiologists have recommended a 30-day quarantine period.

The lack of consensus regarding periods of quarantine reflects the difficulty of interpreting results with small sample sizes. In the case of Ebola virus, previous outbreaks have been limited in many cases to patient populations of just a few hundred. From previous studies, Dr. Haas noted that WHO determined 95% of patients exhibit symptoms of the disease within 21 days after exposure but 5% *do not*. This means that even with a 21-day period of no symptoms, a small number of people will develop the disease and then spread it to others. The current epidemic that has infected more than 20,000 people will provide better statistical data regarding the true period of time that the virus can remain dormant before producing symptoms. However, Dr. Haas concluded that there is "no absolute quarantine time that will provide absolute assurance of no residual risk of contagion."

The probability that Ebola virus disease will spread from an affected to an unaffected area was determined by epidemiologists at the CDC to be directly related to case counts, population density, and the straight-line distance from one area to another. On March 21, 2014, WHO declared that an epidemic of Ebola virus (represented by 49 cases) had begun in Guinea. By August 8, WHO determined that the epidemic had become a condition that constituted a "public health emergency of international concern." Epidemiologists analyzed case count data from the period beginning on March 29, 2014, through August 16, 2104, from three countries (Guinea, Sierra Leone, and Liberia) and determined that the risk of disease spread was significantly affected by spatial relationships between affected and unaffected areas as well as population densities.

© Gordon Swanson/Shutterstock

Additionally, a significant factor affecting disease spread in West Africa occurred from people simply moving from one location to another *without the involvement of air travel*. The Ebola Response Team assigned to the area by WHO determined that the spread of the Ebola virus occurred at an unprecedented rate because of the high degree of interconnections between the three affected countries. Since its first identification in 1976, there have been several outbreaks of Ebola virus in Africa, but they have been relatively contained. Epidemiologists determined that the degree of urbanization in Guinea, Sierra Leone, and Liberia contributed to the rapid expansion of the disease. The principal reason for this was the extensive road connections that exist between international borders and large cities that led to extensive cross-border trafficking. Furthermore, road connections between towns and local villages amplified the speed of transmission. Previous outbreaks had always occurred in remote villages with limited opportunities for spread by travel.

Despite this finding, the Ebola Response Team observed that the most important factor for the rapid spread was the delay of local governments in implementing rigorous control measures. During the 2014 outbreak, the Ebola virus spread to Nigeria but likely did not result in epidemic conditions because of the government's rapid institution of stringent control measures. These included tracing of all contacts with an infected person, followed by their observation for the onset of symptoms, as well as mandatory isolation for those infected. In addition, safe burial practices were imposed, which required embalming the deceased.

▶▶ Outbreak and Control

Epidemiology is the retrospective study of disease occurrences and the determinants of their cause. The beginning of the Ebola virus outbreak garnered virtually no international attention in December 2013. Initially, this was because of geographical factors; beginning in a remote village in Guinea, the outbreak went largely unrecognized by the global community for 3 months except for reports from Médecins Sans Frontières (MSF). MSF, better known in the United States as Doctors Without Borders (DWB), is a humanitarian aid organization largely funded by private donations. Its global mission is the delivery of medical care to people in underdeveloped countries that lack a healthcare infrastructure. In 1999, MSF received the Nobel Peace Prize for its dedicated work to improve health care.

Reports from MSF were provided to officials of the Ministry of Health in Guinea about the spread of the Ebola virus in early 2014. By March 2014, the Ministry of Health of Guinea declared that an epidemic of Ebola virus disease was present in four of its districts. The disease had also spread by this time to the neighboring countries of Liberia and Sierra Leone.

A conflict developed between WHO and the CDC regarding who should handle the epidemic. Although denied by WHO officials, Dr. Thomas Frieden (Director of the CDC) stated that WHO asked the CDC to remove its team of Ebola virus experts from Guinea on the grounds that their help was not necessary. Dr. Frieden has stated that he asked WHO leadership to allow his team to continue its work in Guinea but was asked to leave. Several months later on August 8, 2014, WHO declared the epidemic a public health emergency of international concern with the implication that a **pandemic** was a real possibility.

Accordingly, the CDC moved its Emergency Operations Center to a state of level 1 activation, a condition that also existed during the 2009 pandemic of H1N1 influenza (swine flu) and Hurricane Katrina in 2005. This is the highest level of response and is reserved for critical emergencies only. It allows the CDC to deploy medications, equipment, and people within 6 hours for an international health crisis such as Ebola.

Such a rapid response was needed to contain and control the outbreak. The CDC used experimental epidemiology and computer models to predict the rate of the disease's spread. In October 2014, epidemiologists estimated that the number of Ebola cases in West Africa was doubling every 3 weeks. At that rate, each infected person was passing the disease to two more people.

Strategies to control the outbreak are based on identification and isolation of those infected. The CDC models revealed that to achieve an infection rate of less than one, more treatment centers needed to be built in the affected countries of West Africa. In October 2014, only about 18% of those infected were in treatment centers. The CDC projected that at least 70% of those infected needed to be in treatment centers by the start of 2015 to halt the outbreak's spread. The remaining 30% of those infected could be monitored at home. A typical treatment center is shown in FIGURE 3.10 .

A complicating factor in controlling the epidemic is the burial practices that are common in West Africa. Bodies of the deceased are typically not embalmed in the countries of West Africa involved in the outbreak, and loved ones often wash the bodies of deceased relatives as part of the grieving process. The Ebola virus is believed to survive for several days outside the body of a deceased person, and therefore

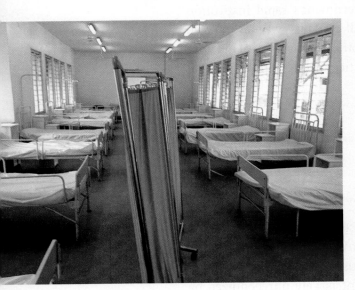

FIGURE 3.10 An example of a treatment center used for Ebola patients in West Africa.

Courtesy of Sally Ezra/CDC

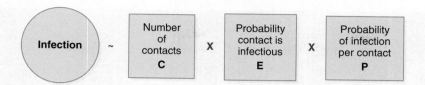

■ **FIGURE 3.11** In any given population, spread of disease is a function of the number of contacts (C), the probability of exposure to an infectious agent (E), and the probability that exposure leads to infection (P).

people became infected from direct physical contact with the deceased due to fluids that remain on the body. Nearly one-third of those infected during the 2014 outbreak were funeral participants.

The institution of safe burials became a strategy to halt the epidemic's spread. Safe burials involved specially trained burial teams who took charge over all aspects of the deceased person's funeral, beginning with removal of the body; relatives were not permitted to touch the body and had to maintain a distance of 15 feet from the deceased. Emphasis is placed on dignity and respect for the deceased throughout the process.

An additional challenge is the difficulty of identifying all of those infected and following up with **contact tracing**. Each person having contact with someone infected must also be identified and contacted to determine whether or not they have developed symptoms. In turn, if any contact has developed symptoms, they must be isolated and their contacts followed, with the pattern repeating whenever a contact develops symptoms. The formula for determining the spread of a disease is shown in **FIGURE 3.11** . In spite of these obstacles, the rate of growth for the disease has slowed when compared to the CDC's earlier projections. Predictions made in September 2014 estimated that without significant intervention to halt the epidemic's spread, about 550,000 cases would occur in West Africa by the middle of January 2015. Control measures have played a significant role in slowing the rate of spread: The total number of cases reported as of April 11, 2015, was just over 25,000, with the total number of deaths over 10,000. The significant difference between the projected numbers and actual cases is a genuine testament to the impact of epidemiology and the dedication of healthcare workers in controlling the rate of spread of one of the world's deadliest infections. To this end, *Time* magazine honored the Ebola fighters as their collective person of the year for 2014.

CRITICAL THINKING QUESTIONS

1. Describe the differences and similarities between an outbreak and an epidemic. When is the term *epidemic* applied to a disease?

2. The rate of infectivity is also known as the reproduction rate for a disease. Why does a higher number for the rate of infectivity indicate that an illness is rapidly spreading within a population? Why does a rate of infectivity less than one indicate that an epidemic is beginning to become under control?

3. Explain the difference between a sign and a symptom.

4. List the three methods used in epidemiological studies of a disease and explain how each is used in studying an outbreak or an epidemic.

5. The CDC has established that the incubation period for the Ebola virus is 21 days. Explain the meaning of the incubation period and why there is debate about extending it beyond 21 days for Ebola.

6. Experimental epidemiology uses several models to predict the outcome of an epidemic. Explain the methods used in the SI, SIR, SIRS, and SEIR models. Which model best fits the Ebola virus? Explain why.

CHAPTER 4

Biotechnology of Ebola

CAROLYN DEHLINGER

Biotechnology is the intersection of science and commerce, where biological systems are engineered in order to synthesize a marketable (and profitable) product. It is part research and part industry, in which scientists from diverse fields of study including genetics, biochemistry, computer programming, molecular biology, and statistics collaborate with project managers, marketing analysts, and sales executives. Living organisms are employed as biofactories, and their cellular machinery is manipulated for a designed purpose. For example, biotechnology has introduced genetic engineering into our agricultural practices. **Genetic engineering** involves the genetic alteration of an organism in order to confer specific trait(s). A commercial crop may be modified to improve its nutrition profile, its tolerance to drought and/or extreme temperatures, or its resistance to invasion by pests.

Alongside agriculture, the healthcare industry is another major sector of the biotechnology industry. A large share of pharmaceuticals on the U.S. market takes advantage of biotechnology in their manufacture. The history of drugs approved

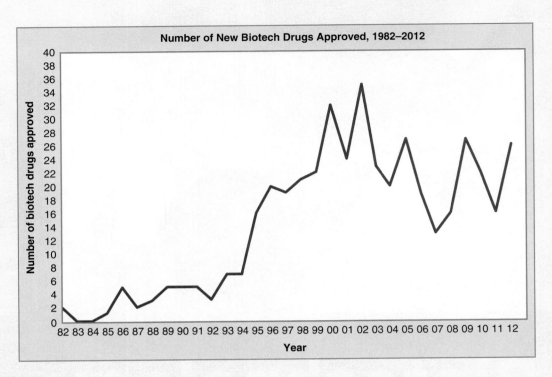

Number of New Biotech Drugs Approved, 1982–2012

■ **FIGURE 4.1** Since the first patent was awarded for a drug produced through biotechnology in 1982, the number approved for market has increased steadily.

by the U.S. Food and Drug Administration (FDA) shows the rise of this class of medicine since the first patent was awarded for a genetically engineered medicine in 1982 (**FIGURE 4.1**). This chapter details the biotechnology related to the Ebola virus. Scientists and healthcare professionals are using biotechnology in order to diagnose Ebola virus infections and in the development of medications for its treatment and vaccines for its prevention.

■ QUESTIONS TO CONSIDER

Some questions to consider as you read this chapter include:

1. What are the existing and experimental diagnostic tests for Ebola virus?

2. What are the desired traits in diagnostic testing for Ebola virus?

3. What are the existing and experimental drug treatments for Ebola virus disease?

4. What prospects exist for development of an Ebola vaccine?

▸▸ Diagnostic Tests

Diagnostic tests are a commonly used tool in medicine aimed at identifying the source of a patient's symptoms. Evaluation of symptoms alone could indicate a number of potential causes. For instance, Ebola virus disease presents in a manner similar to meningitis, typhoid fever, or malaria. In the case of Ebola, a diagnostic test would serve to confirm the presence of the virus in a patient's tissues and/or fluids. While an evaluation of symptoms is beneficial for a preliminary diagnosis, in order to positively identify the nature of an illness a more thorough investigation is required. Diagnostic tests are also used to monitor a patient's progression and response, if any, to treatment(s). Due to Ebola virus's classification as a Biosafety Level 4 infectious agent (BSL-4), it should be noted that all diagnostics must be performed under the strictest protective measures. The BSL-4 designation is given to the most virulent pathogens by the **Centers for Disease Control and Prevention (CDC)**. They are classified by their high fatality rates and lack of vaccines and/or treatments. Many viral hemorrhagic fever pathogens are classified at the BSL-4 level, such as Lassa fever, Nipah virus, and Ebola's close relative, *Marburgvirus*.

Although the first Ebola outbreak took place nearly 40 years ago, the medical community was unprepared for the current outbreak as far as diagnostic testing is concerned. One reason for this is lack of resources in affected countries. Diagnostics are critical not just for patient monitoring but also for support in contact tracing. Transportation limitations throughout West Africa make placement of diagnostic centers near Ebola Treatment Units (ETUs) most ideal. Distance between diagnostic centers and ETUs is positively correlated to the amount of time between sample collection and results reporting. The greater the distance, the longer it takes to find out the results of diagnostic tests. At present, the reality does not reflect the recommendation to place these facilities close to each other. Time lags are a considerable hindrance in rapid diagnoses. Current conditions clock the time for diagnostic testing, from collection to results, anywhere from 4 hours to 9 days.

Additionally, the following diagnostic procedures are by no means "standard" or in widespread, systematic use. The **World Health Organization (WHO)** has established a prequalification process in order to evaluate the potential application of diagnostic methods such as those described in the sections that follow. The prescribed Expedited Product Review and Emergency Quality Assurance Mechanism will allow WHO to disseminate appropriate information within the affected countries. The Foundation for Innovative New Diagnostics, or FIND, is working with WHO and Médecins Sans Frontières (MSF, the French arm of Doctors Without Borders [DWB], which has contributed significantly to the outbreak response) in this regard. This collaboration seeks the identification, development, and implementation of a rapid diagnostic test that requires perhaps a few drops of blood instead of a test tube vial and that can be performed and read without the need for sending it off to a laboratory. As of December 2014, a 5-month plan has been in place to reach this goal.

Antigen-Capture ELISA: Detecting Ebola Antigens

One current diagnostic test for Ebola detects the presence of antigens for the virus, often from a patient blood sample. **Antibodies** are proteins produced by the immune system in response to foreign proteins known as **antigens**. The immune system makes a new type of antibody with each unique antigen that enters the body. Therefore, it is possible to develop a diagnostic for the antigen specific to an Ebola virus.

Antigen-capture **enzyme-linked immunosorbent assay (ELISA)** is one technique used to identify the presence of a specific antigen. ELISA is a qualitative test, with either a positive or negative result. Ebola is studded with glycoprotein antibodies, and these are used in diagnosing infections of the virus. In the Ebola-specific test, ELISA makes use of a plate of wells coated with the glycoprotein antibodies that are linked to enzymes. The glycoprotein antibodies will bind to Ebola antigens, if present in the sample being tested, while the enzymes cause a color change indicative of a positive result (**FIGURE 4.2**). The manner in which the glycoprotein antibodies were first collected is discussed in the Monoclonal Antibodies section later in this chapter.

Researchers at Tulane University School of Medicine are working in collaboration with the U.S. pharmaceutical company Corgenix on a cost-effective and quick version of antigen-capture ELISA. Corgenix is already established within the *in vitro* diagnostic market and offers various ELISA kits across the world, in addition to offering other detection methods. In a method similar to a home pregnancy test, a finger prick draws a drop of blood onto a testing strip. The strip is then placed in a testing

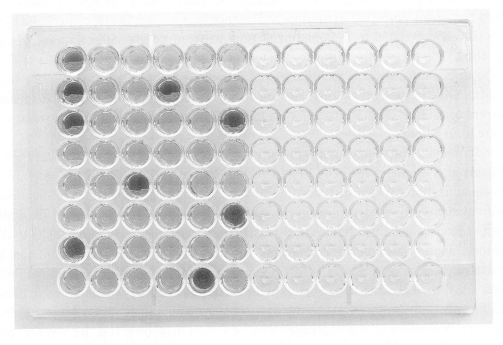

FIGURE 4.2 The color change (red) on an ELISA plate indicates a positive match. ELISA is a qualitative diagnostic test.

TABLE 4.1	World Health Organization Emergency Quality Assurance Mechanism Antigen-Capture ELISA Candidates
Testing Candidate	**Manufacturer**
Ebola eZYSCREEN Rapid Test	CEA
Ebola Antigen Detection	Chembio Diagnostics
ReEBOV™ Antigen Rapid Test (*Zaire ebolavirus* Antigen Detection)	Corgenix
Biocredit Ebola Ag	RapiGEN Inc.
DEDIATEST-Ebola	Senova GmbH
ABICAP®-Rapid-Ebola	Senova GmbH

Source: Modified from World Health Organization. Emergency QA Mechanism of IVDs for Ebola Virus Disease. *In vitro* diagnostics and laboratory technology. January 9, 2015. http://www.who.int/diagnostics_laboratory/procurement/purchasing/en/. Copyright 2015 World Health Organization.

solution containing Ebola antibodies. Results appear in approximately 15 minutes in the form of a solid line indicative of a positive result. However, with antigen-capture ELISA, a person being tested is unlikely to build up a detectable level of antigens soon after infection. This diagnostic method is better suited to testing individuals experiencing symptoms, thus it would not be as effective in preemptive screenings in large populations such as in airports. A complete list of antigen-capture ELISA products currently being evaluated by WHO under its Emergency Quality Assurance Mechanism initiative is summarized in **TABLE 4.1**.

Reverse Transcription Polymerase Chain Reaction

Another means of diagnosing Ebola virus infections is the reverse transcription polymerase chain reaction (RT-PCR). This method identifies ribonucleic acid (RNA) of Ebola virus within infected cells of a patient. RT-PCR is a modified form of the polymerase chain reaction. The PCR method as it was originally conceived amplifies small quantities of deoxyribonucleic acid (DNA) into exponentially larger amounts. An original DNA molecule is used as a template to produce millions of identical copies in a piece of machinery no larger than a toaster oven. Invented in 1985 by Kary Mullis, then an employee of the large biotechnology company Cetus Corporation, it relies on the concept of exponential growth. One DNA molecule is copied into two, which are then copied to make four, and so on. The technology won Mullis the 1993 Nobel Prize for Chemistry. The process is highlighted in **FIGURE 4.3**.

With the RT-PCR, DNA is amplified from an RNA template rather than a DNA template. The enzyme reverse transcriptase, also known as RNA-dependent DNA polymerase,

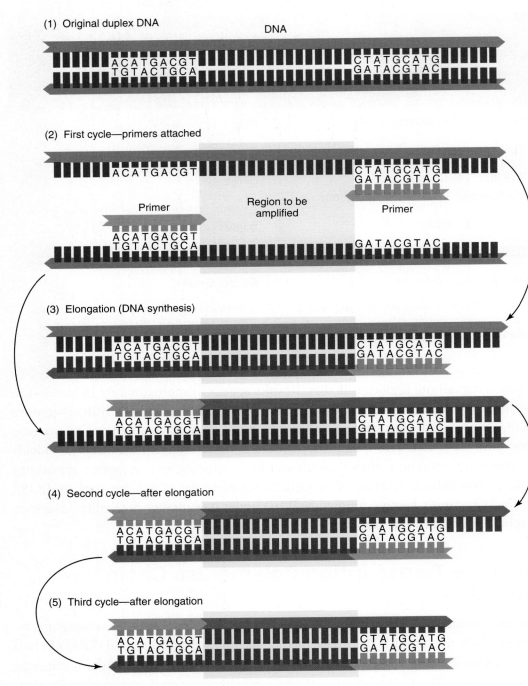

(a)

FIGURE 4.3 The polymerase chain reaction consists of three steps. (a) DNA is denatured, primers are annealed, and elongation occurs via complementary base pairing. (*continues*)

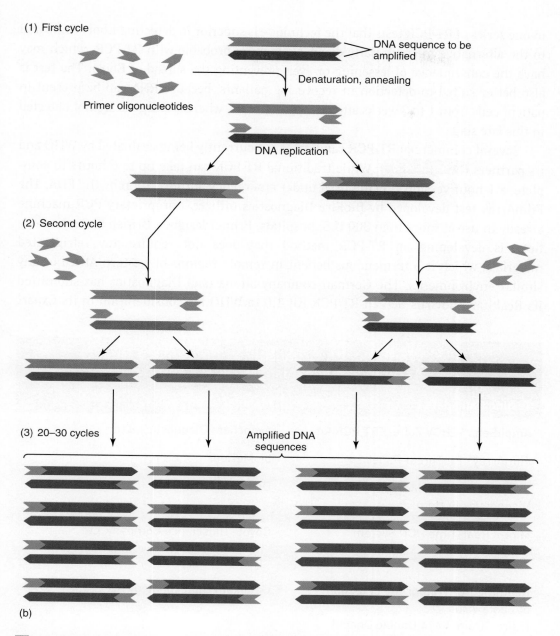

(1) First cycle

DNA sequence to be amplified

Primer oligonucleotides

Denaturation, annealing

DNA replication

(2) Second cycle

(3) 20–30 cycles

Amplified DNA sequences

(b)

FIGURE 4.3 (*continued*) (b) PCR amplifies DNA in an exponential fashion.

is a naturally occurring enzyme crucial to the retrovirus lifecycle. Reverse transcriptase enzymes are introduced into the process as it synthesizes DNA from RNA. Because the Ebola viruses possess RNA as its genetic material, this method is capable of identifying Ebola RNA within an infected patient. The amplified DNA is known as complementary DNA or cDNA. Fluorescent probes complementary to target Ebola RNA sequences are utilized in the process in order to scan synthesized cDNA and positively identify the viral infection. The RT-PCR appears to be much more sensitive in its diagnostic capabilities as compared to antigen-capture ELISA. This is likely because with RT-PCR even minute quantities of Ebola RNA will be recognized. Preliminary investigation has found in one series of RT-PCR tests that the technique is superior in detecting Ebola compared to the alternative method. Early diagnosis is more probable with RT-PCR, which may have the consequence of lessening or even preventing the spread of Ebola. The test is also better suited to detection in recovering patients, because RNA can be present in patient cells from 1 to 3 weeks after symptoms cease, whereas antigens are not detected in this late stage.

Several commercial RT-PCR candidates are currently being evaluated by WHO and its partners (TABLE 4.2). While traditional RT-PCR can take up to 6 hours to complete, a 1-hour version is being used under emergency authorization by the FDA. The FilmArray test developed by BioFire Diagnostics utilizes a proprietary PCR machine already in use at more than 300 U.S. hospitals. Primerdesign, a British biotechnology firm, is developing an RT-PCR method that does not require any refrigerated materials, which is a tremendous benefit in remote regions of western Africa or any similar environments. The German company altona (sic) Diagnostics has submitted its RealStar Filovirus Screen RT-PCR Kit 1.0 to WHO for consideration in its Expert

TABLE 4.2	World Health Organization Emergency Quality Assurance Mechanism RT-PCR Candidates
Testing Candidate	**Manufacturer**
AmpliSens® EBOV Zaire-FRT PCR kit	Avonchem Diagnostics, Ltd
FilmArray® BioThreat Panel	bioMérieux
ExiStation™ EBOV Molecular Diagnostics System	Bioneer
Mini-8 Real-Time PCR System	Coyote Bioscience Company, Ltd
Epistem Genedrive®	Epistem Plc
Realtime PCR-Kit for the detection of Ebola Virus (Zaire Strain; K414 Double Check)	Genekam Biotechnology AG
QuickGene-Mini80	Kurabo Industries, Ltd
Ebola Virus (Zaire 2014) Assay and Control Set	Life Sciences Solutions
Genesig® Easy DNA/RNA Extraction Kit + Genesig® Advanced EBOV_2014 qPCR Detection Kit	PCRmax, Ltd and Primerdesign, Ltd
Life River	Shanghai ZJ Bio-Tech Company, Limited
LightMix® Ebola Zaire rRT-PCR Test	TIB MOLBIOL Syntheselabor GmbH

Source: Modified from World Health Organization. Emergency QA Mechanism of IVDs for Ebola Virus Disease. *In vitro* diagnostics and laboratory technology. January 9, 2015. http://www.who.int/diagnostics_laboratory/procurement/purchasing/en/. Copyright 2015 World Health Organization.

Product Review described earlier. This kit is a full-spectrum diagnostic that detects all filoviruses pathogenic to humans as well as *Reston ebolavirus*, which does not affect humans. Because the kit contains unique fluorescent probes for each filovirus, including Marburg and Ebola, it can differentiate between pathogens and specify what infection, if any, is present.

▸▸ Drug Treatments

In September 2014, a meeting took place in Geneva, Switzerland, at the WHO headquarters. Six teams called Expert Working Groups were tasked with identifying unregistered or unapproved drugs that may hold promise in combating Ebola. They presented the information they collected at this WHO meeting in Geneva. The Expert Working Groups were composed of:

- Virologists and vaccinologists
- Pharmacologists and clinicians
- Policymakers and regulators
- Research funders
- Pharmaceutical industrialists
- Ethicists and esteemed members of society

Attended by health officials and pharmaceutical industry representatives, the meeting sought to identify the leading candidates for effective treatment of Ebola virus infections in humans. The main questions the meeting sought to answer revolved around the safety and efficacy of treatments, the ease of rapid development in bringing a potential treatment or vaccine from the laboratory into the field, and the logistics of scaling up production to reach demand. A summary of these Ebola virus treatments is presented in the sections that follow.

Monoclonal Antibodies

Monoclonal antibodies (MAbs) comprise a single type of antibody for treatment (or identification) of a particular genetic disorder or disease. The first procedure to synthesize MAbs for clinical use was developed in 1975. Murine rodents such as mice or rats are injected with a purified antigen that is complementary to the antibody under production. Time is given for the murine immune system to respond, which results in biosynthesis of the desired antibodies. After several weeks, the animal's spleen is removed. The spleen is the location of B lymphocytes that produce antibodies, so it is the richest source available. The B lymphocytes are cultured in the presence of cancerous myeloma cells, which are immortal and will grow in culture indefinitely. Scientists can create the right environmental conditions in culture to induce fusion of B lymphocytes to the myeloma cells, resulting in hybrid cells known as **hybridomas** (FIGURE 4.4). The fusion can be encouraged by electric shock, introduction of a virus, or through treatment with polyethylene glycol.

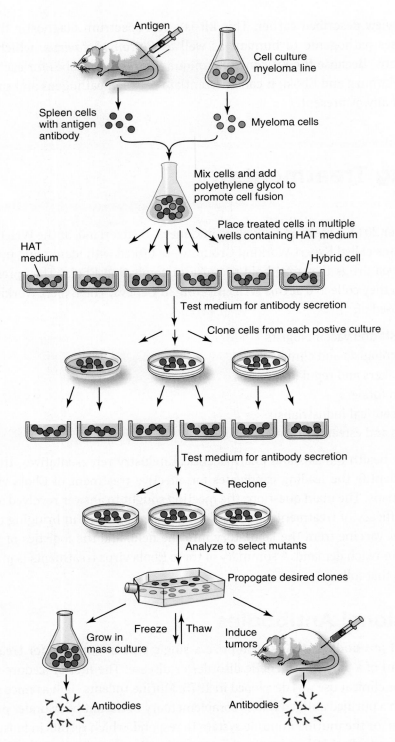

FIGURE 4.4 An overview of how a hybridoma cell line is established in the laboratory.

Source: Adapted from Milstein, C. (1980). Monoclonal antibodies. *Scientific American*, 241, 66–74.

Hybridoma cultures are living factories for MAb production, because the B lymphocyte portion of the hybrid is encoded for the antibody. As long as environmental conditions are maintained, the myeloma portion keeps the culture dividing ad infinitum. Hybridomas can be fermented in batches or continuous cultures within industrial bioreactors to increase production.

Mapp Biopharmaceuticals has developed an MAb treatment for Ebola virus known as **ZMapp**. Due to the intensive manufacturing process, the drug is being produced in batches and shipped to western Africa as it becomes available. Once in country, the drug is involved in an ongoing clinical trial. ZMapp was well publicized in fall 2014 as the drug administered to U.S. doctor Kent Brantly after he was air-lifted from the Democratic Republic of Congo to Emory University Hospital in Atlanta, Georgia. While it saved Dr. Brantly's life, ZMapp was ineffective in other patients. The Public Health Agency of Canada tested ZMapp in 21 infected nonhuman primates with success. Of the 21 infected, 18 were treated with ZMapp up to 5 days following the onset of symptoms. All of the treated subjects survived their infections.

ZMapp is administered as a single, three-dose course of treatment. Interestingly, with ZMapp another cell type was engineered in order to synthesize MAbs. ZMapp contains three separate *Ebolavirus* MAbs that are manufactured by engineered tobacco plants (*Nicotiana benthamiana*). Production of a drug via the cultivation of plants is an example of **biopharming**. The tobacco plants contain genes originally constructed in murine hybridomas, which enables the production of cells containing Ebola MAbs and human antigens to Ebola. These antigens prevent rejection of ZMapp-produced Ebola antibodies in human patients. Because the genes used in synthesizing ZMapp are part human and part rodent, the drug constituents are termed **chimeric antibodies**. The antibodies defend against glycoproteins on the outer coat of Ebola viruses.

The company Kentucky Bioprocessing assists Mapp Biopharmaceuticals in ZMapp manufacture by cultivating these genetically engineered plants. The entire manufacturing process is painstakingly slow, contributing to low supply levels that were exacerbated by scarce demand, at least until the 2014 outbreaks. Because Kentucky Bioprocessing can produce on average just 12 doses per month, additional companies specializing in MAb manufacture are needed.

Small-Molecule Pharmaceuticals

In 1982, Eli Lilly and Company placed a human insulin medication on the market to treat patients with type 1 diabetes. The patent for this drug, known as Humulin®, was granted based on the novelty of its manufacturing process. **Recombinant DNA technology** was conceived as a genetic engineering technique by Stanley Cohen in 1975. The method uses bacterial plasmids as vectors in order to introduce foreign DNA, known as a **transgene**, into the genome of a bacterial host cell. The bacterial cell then contains a new combination of DNA that is part its own and part foreign (i.e., recombinant DNA), thus making it a transgenic or **genetically modified organism (GMO)**. The transgenic bacteria is then cultured and allowed to express its inserted transgene. Using this process, a desired product is synthesized within large cultures of bacterial cells,

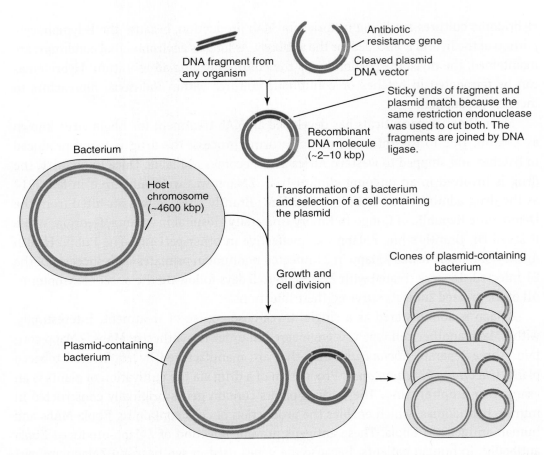

FIGURE 4.5 In 1975, Stanley Cohen invented the technique that would become known as recombinant DNA technology, or in his words a "molecular cloning procedure."

which are then harvested for the product's extraction. In 1976, Cohen cofounded the first biotechnology company to be listed on the New York Stock Exchange, Genentech. It was Genentech that developed human insulin synthesized through recombinant DNA technology in 1978. Once licensed and marketed by Eli Lilly and Company, it became known as Humulin. The process of **molecular cloning**, which utilizes recombinant DNA technology to "clone" myriad copies of a molecule such as human insulin, is detailed in **FIGURE 4.5**. Drugs manufactured through this process represent small-molecule pharmaceuticals, such as the following medications under investigation for the treatment of Ebola virus infections.

Brincidofovir

Brincidofovir is an antiviral drug owned by Chimerix, Inc. (Durham, NC). Interestingly, brincidofovir came into being due to national efforts to combat bioterrorism. A research team at the University of California, San Diego (UCSD) led by Dr. Karl Hostetler created the drug, then known as CMX001, in a project funded by the National Institute of Allergy and Infectious Diseases (NIAID), a division of the **National Institutes of Health**. The goal of the project was to improve upon the Gilead Sciences antiviral medication

TABLE 4.3	Primary Considerations in the Development of Stockpiled Emergency Medications
Desired Trait	**Benefit(s)**
Active in pill or capsule form	Long-term storage; ease in use if ever necessary
Stable formulation	Long-term storage
Minimal side effects	Reduces need for secondary medical treatment(s) if ever necessary
Safe for compromised immune systems	Increased participation if ever necessary

Source: Data from Kroll, D. Why everyone has a stake in the Chimerix drug offered to Josh Hardy. March 12, 2014. http://www.forbes.com/sites/davidkroll/2014/03/12/why-everyone-has-a-stake-in-the-chimerix-drug-offered-to-josh-hardy-cmx001-brincidofovir/3/

Vistide® (generic name cidofovir), which was effective against a broad range of viruses but caused significant side effects, with kidney damage being of primary concern. Cidofovir has activity on all families of double-stranded DNA viruses known to infect humans, including herpesviruses, smallpox variola virus, adenoviruses, cytomegaloviruses, and polyomaviruses. The anti-bioterrorism aspect of the project focused on how cidofovir might be improved to act as a stockpiled response in the event of a smallpox attack. For a drug to serve well in a massive, nationwide bioterrorism response, several key characteristics are desired. Summarized in **TABLE 4.3**, these traits were also major goals of the UCSD project.

Hostetler's team hypothesized cidofovir might be improved by attaching the drug to a class of lipid molecules known as fatty acids. The addition of lipids was shown to increase the uptake of the medication into target cells. In this case, target cells included those of the intestinal tract, because the medication would be taken orally and would have to withstand passage through the digestive system. The specific lipid molecule used is chemically similar to a naturally occurring lipid found within the human body known as lysophosphatidylcholine (LPC). Use of LPC allows human cells to recognize CMX001 as "normal," which improved the uptake of the drug by 50 times compared to Vistide. The increase in uptake resulted in a 40–400 times improvement in antiviral activity.

Tagging the cidofovir drug with LPC created what is called a **prodrug**. CMX001 behaves as a payload of cidofovir, whereupon entering a cell the lipid molecule is removed, thereby allowing the active ingredient to perform its antiviral function. Cidofovir prevents DNA replication, a key step in the viral lifecycle (**FIGURE 4.6**). Without DNA replication, a virus is unable to make copies of itself inside a host cell, thus preventing its ability to reproduce and infect other cells.

Conveniently, CMX001 has been shown to target viral DNA synthesis enzymes and does not appear to greatly affect human DNA synthesis enzymes. In other words, the drug is effective in stopping viruses from reproducing but avoids getting in the way

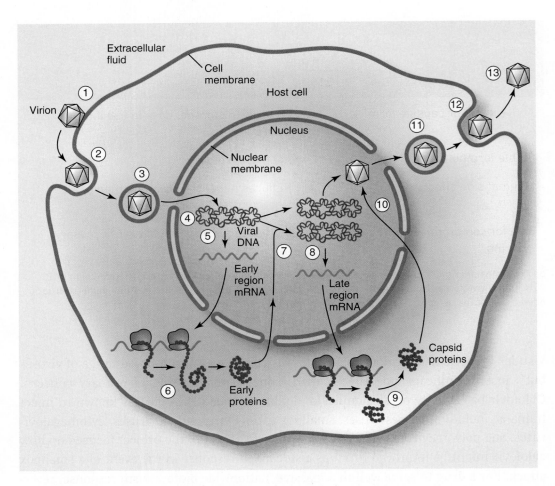

FIGURE 4.6 This schematic details the replication cycle of a polyomavirus. Cidofovir has been shown to interfere with this life cycle.

Source: Adapted from Knipe, D. M., & Howley, P. (1996). *Fundamental virology* (3rd ed.). Philadelphia: Lippincott Williams & Wilkins

of normal cell growth. The exact cause of this "happy accident" is unknown, but the effect is well appreciated. It is uncertain how human cells remain relatively unaffected by cidofovir and are able to continue normal DNA replication and cell growth even in the presence of the drug. What is understood is the means by which viruses are affected by CMX001. Upon entering a host cell, two phosphates are added to cidofovir, creating a cidofovir diphosphate molecule. Viral DNA synthesis enzymes preferentially utilize cidofovir diphosphate, because it is chemically similar to a building block of DNA known as cytidine.

Another advantage of the LPC component is reduced kidney damage. Cidofovir alone interferes with a protein pump in kidney cell membranes, causing the drug to accumulate within kidney cells. When attached to LPC, accumulation is prevented. Upon patent application of CMX001, the biotechnology was licensed to Chimerix by UCSD. Chimerix has engaged in clinical trials of brincidofovir for the treatment of adenoviruses and cytomegaloviruses, with a focus on cancer patients receiving hematopoietic cell transplants as well as whole organ transplant patients with compromised immune systems.

Although originally developed for the treatment of double-stranded DNA viruses, the drug later showed some promise in combating Ebola derived from an isolate of the first outbreak in 1976. Although Ebola is a single-stranded RNA virus, the CDC included brincidofovir in a broad screening of antiviral medications in an emergency search for Ebola treatments in fall 2014. The screening treated cell cultures infected with the Ebola isolate with a host of antiviral medications in the hopes of finding effective treatments already in existence. Brincidofovir was then used as an emergency treatment for at least two infected patients transported to the United States, including Thomas Eric Duncan, the Liberian man who traveled to Dallas, Texas, and Ashoka Mukpo, an NBC News cameraman who was treated in Omaha, Nebraska,

© Alex011973/Shutterstock

at the University of Nebraska Medical Center's Biocontainment Unit. Mukpo contracted the Ebola virus while on assignment in Liberia. The Dallas patient did not respond to brincidofovir and died on October 8, 2014. The Nebraska patient survived and was declared free of the virus in late October 2014.

The University of Oxford and the International Severe Acute Respiratory and Emerging Infection Consortium (ISARIC) oversaw a brief trial investigation of brincidofovir at MSF's ELWA 3 Ebola Management Centre in Monrovia, Liberia, during the month of January 2015 (ELWA stands for Eternal Love Winning Africa). The study was abruptly halted by Chimerix on January 30, 2015. Citing a declining rate of incidence in the hardest hit West African countries, namely Liberia, Guinea, and Sierra Leone, the company announced they would be ceasing further study of the medication as a treatment for Ebola. Rather, Chimerix planned to return its focus to the drug's efficacy in treating adenovirus and cytomegalovirus. During such a brief interval of investigation, Chimerix reported that only a handful of patients had participated in the Liberia trial. Oxford researchers had planned to enroll up to 140 patients, who otherwise would have received doses of brincidofovir over a 2-week period. The experimental design of the study intended for the survival rates of enrolled participants to be compared with historical data of the current and previous outbreaks. In the span of a few weeks, Chimerix initiated trials by rushing stockpiles of the medication to Liberia on U.S. military aircraft only to reverse their decision after talks with the FDA.

BCX-4430

BCX-4430 is a small-molecule drug developed by BioCryst Pharmaceuticals. It has shown promise in fighting both Ebola and Marburg viruses in rodents and Marburg virus in nonhuman primates. BCX-4430 acts by inhibiting the RNA-dependent RNA polymerase enzyme, thereby preventing replication of the virus. More specifically, the drug binds to the active site of the polymerase enzyme and becomes incorporated into newly synthesized RNA. This causes the replication of the viral RNA to terminate prematurely, rendering malformed and nonfunctional virions. As with CMX-001, NIAID is involved in the funding of BCX-4430 development. NIAID initially engaged in the study with BioCryst in order to investigate Marburg virus treatments, with a secondary consideration of other filoviruses such as Ebola. Since the 2014 outbreaks, the project has shifted its focus slightly to entail the development of a wide-spectrum drug capable of treating multiple viral hemorrhagic fevers. In addition to intravenous and intramuscular versions, BCX-4430 is manufactured as a stable pill form, making its storage, dissemination, and use relatively simple.

RNAi

RNA interference (RNAi) is a form of gene silencing that can be used as a method of gene therapy. As the name implies, RNA interference uses RNA as a silencing molecule that interferes with normal gene expression (FIGURE 4.7). RNAi uses an enzyme called Dicer to cut RNA into small interfering RNAs (siRNAs). Small interfering RNAs bind to a quaternary protein complex called the RNA-induced silencing complex

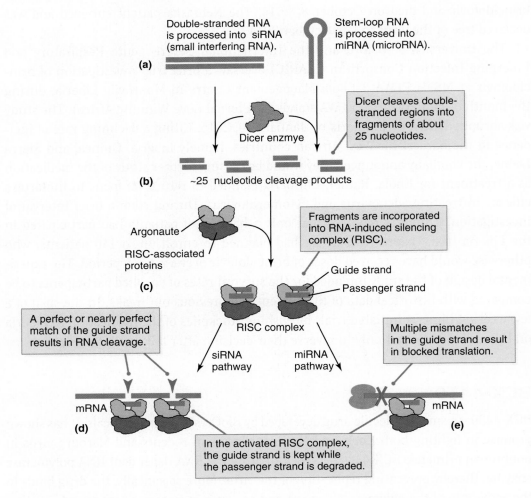

FIGURE 4.7 The modes of action in RNAi technology. RNA interference may utilize small interfering RNA molecules (left) or microRNA (right).

(RISC). The RISC delivers siRNAs to their target mRNA, in this case that of the replicating virus, and both types of RNA bind through complementary base pairing. RISC then degrades the triplex of RNA, preventing translation and the synthesis of the infectious virions. RNA interference may also be achieved with native microRNAs (miRNAs) instead of introduced siRNAs. The mechanism of gene silencing inside the cell is similar when using miRNAs, but because miRNAs already exist in a cell, it is only necessary to stimulate their activity. RNA interference has proven successful in research studies for the suppression of more than a dozen viruses, so it will likely also be used in the future for treatment and/or prevention of viral infections. Examples of viruses suppressed in culture via RNAi include hepatitis C, human papillomavirus, HIV, and, now, filoviruses like Ebola.

TKM-Ebola

Canadian company Tekmira Pharmaceuticals has developed an RNAi therapy for filoviruses with funding by the Defense Threat Reduction Agency of the U.S. Department of Defense. At the University of Texas, Tekmira tested its RNAi therapy in nonhuman primates infected with Marburg virus; all infected subjects treated with the drug survived. Much like with ZMapp studies, this treatment was also effective even days after the onset of symptoms. An additional study with nonhuman primates conducted at Boston University in collaboration with the U.S. Army Medical Research Institute of Infectious Diseases (USAMRIID) had similar results with nonhuman primates, only this time with *Zaire ebolavirus*, the same strain of Ebola involved with the current outbreaks.

What makes the Tekmira system unique compared to RNAi as described earlier is its modified delivery system, known as lipid nanoparticle (LNP) technology. Tekmira's therapeutic agent uses siRNA to interfere with the action of three of the seven total Ebola genes. In order to effectively deliver siRNA into a patient's cells where it can then interfere with virus replication, Tekmira envelops the small interfering RNAs in LNPs. This layer of lipid molecules prevents the patient's immune system from attacking the therapeutic agent, giving the drug time to enter target cells where it will be most effective. Target cells include immune system cells and liver cells. Viruses like Ebola do considerable damage to liver cells, rendering them ineffective at clotting blood, which leads to the symptomatic hemorrhaging associated with viral hemorrhagic fevers.

In January 2014, TKM-Ebola, as the treatment is called, was used in a human clinical trial that concluded the following May. This trial was conducted in healthy human volunteers in order to assess potential side effects. In March of the same year, the drug received a Fast Track designation from the FDA designed to allow a shorter timeline from clinical trials to market availability. Trials are now limited to Ebola-infected patients or those suspected to be infected. To date, a small number of patients have received TKM-Ebola, according to Tekmira's website. Researchers are now contemplating a drug cocktail of sorts that would be effective in treating the whole filovirus family. For example, the nanoparticle delivery system could be engineered to contain not only siRNAs but also MAbs such as those present in ZMapp.

AVI Antisense Therapies

The therapeutic agent known as AVI-7537 was developed by Sarepta under an agreement with the U.S. Department of Defense. While TKM-Ebola silences three Ebola genes, AVI-7537 interferes with only one of those: the VP24 gene. Sarepta also created a separate version of the drug for Ebola called AVI-7539, which the company has tested in combination with AVI-7537, creating a separate entity known as AVI-6002. All three have been tested in nonhuman primates against *Zaire ebolavirus*, while a separate cocktail called AVI-6003 (composed of AVI-7287 and AVI-7288) has been used similarly in treatment of Marburg virus. AVI-6003 interferes with the NP gene of Marburg virus.

Sarepta took a different approach for its delivery system than did Tekmira. Instead of lipid nanoparticles containing small interfering RNA molecules, Sarepta uses phosphorodiamidate morpholino oligomers (PMOs) as its mode of action. The company developed a proprietary form known as PMO*plus* that contains positively charged linkages. PMO behaves as **antisense RNA**, meaning it binds to viral RNA and prevents it from being copied (FIGURE 4.8). Thus, the viral lifecycle is inhibited. PMO*plus* appears to be especially stable compared to other forms in addition to improving binding to viral RNA and, therefore, its efficacy in preventing replication.

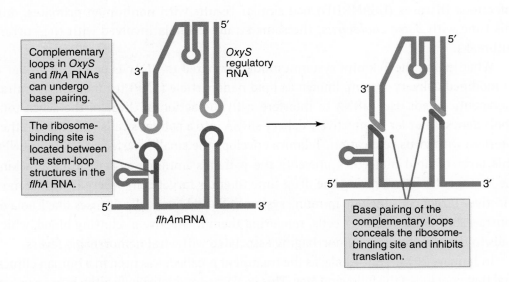

FIGURE 4.8 Sarepta employs phosphorodiamidate morpholino oligomers (PMOs) as a form of antisense RNA technology. Antisense RNA blocks the expression of filovirus genes, preventing virion replication.

Source: Adapted from Altuvia S., & Wagner, E. G. H. (2000). Switching on and off with RNA. *Proceeding of the National Academy of Sciences of the United States of America, 97,* 9824–9826.

▸▸ Vaccine Development

Vaccines introduce foreign infectious agents into the human body in order to stimulate an immune response. Vaccines are so named because the first vaccine contained cowpox derivatives (*vacca* is the Latin word for cow). Vaccination for certain diseases is a routine medical procedure. There are four major categories of vaccines. They are grouped based upon their mode of action (TABLE 4.4). Many factors must be considered when developing a new vaccine. The characteristics of an ideal vaccine are described in TABLE 4.5.

In late January through early February 2015, the number of reported new cases of Ebola patients in Guinea nearly doubled. Sixty-five new cases were reported as of February 8, 2015, whereas the week prior 39 new cases were reported, according to WHO. Responding to this reversal in an overall downward trend for incidence of infection, the government of Guinea called for the expansion of administering experimental drugs such as favipiravir (Avigan). Avigan was approved in Japan as an anti-influenza

TABLE 4.4	Types of Vaccines	
Type of Vaccine	**Active Component**	**Examples**
Attenuated	Live but weakened bacteria or viruses	• Polio (Sabin vaccine) • MMR • Chickenpox • Cholera • Tuberculosis
Inactivated (also known as killed)	Dead bacteria or viruses	• Polio (Salk vaccine) • Rabies • DPT • Influenza
Subunit	Portions of bacterial or viral structures, such as proteins or lipids	• Hepatitis B • Anthrax • Tetanus • Meningitis
DNA	Bacterial or viral DNA	• West Nile virus • Canine melanoma

TABLE 4.5	Characteristics of an Ideal Vaccine
Produces a good humoral, cell-mediated, and local immune response, similar to natural infection, in a single dose.	
Elicits protections against clinical disease and reinfection.	
Provides protection for several years, preferably a lifetime.	
Results in minimal immediate adverse effects or mild disease with no delayed effects that predispose to other diseases.	
Induced immunity confers protections to multiple strains of organisms.	
Can be administered simply in a form that is practically, culturally, and ethically acceptable to the target population.	
Vaccine preparations do not require special handling (e.g., a cold chain).	
Does not interfere significantly with the immune response to other vaccines given simultaneously.	
Costs and benefits associated with receiving the vaccine clearly outweigh the costs and risks associated with natural infection.	

medication. Toyama Chemical Company, a subsidiary of Fujifilm Holdings Company, developed the drug now being tested in Guinea. Initial trials began in mid-December 2014 by both Guinean and French medical teams, represented by the French Institute of Health and Medical Research (INSERM), the French Red Cross, and MSF. Currently, the medication is available at designated ETUs only, not in hospitals. While representatives of the Avigan trials have gone on record stating positive results in patients treated with the medication, no hard data have been released as of this writing.

In Liberia, another country greatly affected by the Ebola outbreaks, vaccine trials are ongoing. cAd3-EBOZ is the name of a vaccine candidate developed in the United States as the result of a partnership between pharmaceutical conglomerate GlaxoSmithKline and

NIAID. This vaccine uses a chimpanzee cold pathogen to deliver genetic material from the *Zaire* strain of Ebola virus. Liberia and NIAID are cooperating in vaccine studies as the organization Partnership for Research on Ebola Vaccines in Liberia (PREVAIL).

PREVAIL is also investigating another vaccine candidate known as VSV-ZEBOV. VSV represents the livestock pathogen contained within the vaccine. Known as vesicular stomatitis virus (VSV), the virus commonly infects cows, horses, pigs, and, to a lesser extent, sheep and goats. Licensed to a subsidiary of the Iowa biotechnology firm NewLink Genetics called BioProtection Systems Corporation and later to pharmaceutical conglomerate Merck, this vaccine is administered through the Public Health Agency of Canada. VSV-ZEBOV contains a genetically modified form of VSV that carries a *Zaire ebolavirus* (ZEBOV) transgene. A transgene is a gene of foreign origin inserted into the genome of another species in the process of genetic modification.

Both vaccines are entering second-phase trials on healthy participants and at-risk volunteers. PREVAIL has identified at-risk populations to include healthcare workers, Ebola containment workers, residents of hotspot area villages, and burial crews handling deceased Ebola patients. Redemption Hospital in Monrovia, Liberia, seeks to administer vaccines to some 27,000 individuals. The double-blind study will divide participants into one of three groups: a **placebo** (or control) group, a second group treated with cAd3-EBOZ vaccine, and the last group treated with VSV-ZEBOV vaccine. The **double-blind** nature of the study ensures neither the researchers administering the study nor the participants will be aware of which treatment each volunteer receives.

CRITICAL THINKING QUESTIONS

1. Are there any moral or ethical concerns regarding the testing and treatment of Ebola? What do they include?

2. The experimental treatment ZMapp saved the life of Dr. Kent Brantly, yet other patients receiving the same medication did not survive. What might account for this discrepancy?

3. What possible circumstances may have limited the research of Ebola diagnostic tests, treatments, and vaccines prior to the most recent outbreaks?

4. Evaluate the pros and cons for: (a) each experimental treatment and (b) for each vaccine in development.

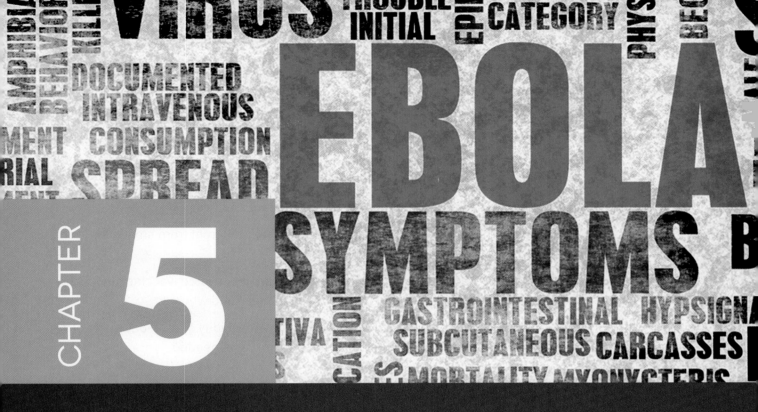

CHAPTER

5

Response to the Ebola Epidemic

George Ealy

The current epidemic of Ebola virus is the 23rd since the disease was first recognized in 1976. To date, it has infected more than 23,000 people, which is an infection rate of roughly 10 times higher than all previous 22 outbreaks. What is different about this epidemic? Why have so many people become infected? Why was this epidemic not contained in its early stages as with previous outbreaks? How did this epidemic spread so quickly to involve so many areas? And perhaps most importantly, how can public health agencies prepare for the next outbreak to prevent an even worse outcome? Many of the answers to these questions are directly related to the international community's *response* to the Ebola epidemic.

■ QUESTIONS TO CONSIDER

Some questions to consider as you read this chapter include:

1. Why did the World Health Organization (WHO) wait several months before declaring that an epidemic of Ebola virus was occurring?

2. What role did social media play in the 2014 Ebola outbreak?

3. What are monoclonal antibodies, and how are they different from small interfering RNA molecules?

4. What is the essential difference between an inactivated vaccine and a live attenuated vaccine? Which is safer for the patient and why?

5. Explain why some epidemiologists challenge the validity of a 21-day quarantine period and think that a longer time for quarantine is needed.

▸▸ Governmental Policies

World Health Organization Response

The World Health Organization (WHO) (**FIGURE 5.1**) announced in April 2015 that it had been slow in responding to early reports of the widespread outbreak. Speaking in Geneva on March 22, 2014, Gregory Hartl, a spokesman for WHO stated, "This [Ebola outbreak] is relatively small still. The biggest outbreaks have been over 400 cases." His remarks were made on the same day that WHO posted news on its website that the Ministry of Health of Guinea had reported the occurrence of a "rapidly evolving outbreak of Ebola virus disease (EVD) in forested areas of southeastern Guinea." However, the international medical charity, Médecins Sans Frontières (MSF), known in the United States as Doctors Without Borders, warned WHO in early April 2014 that the Ebola epidemic was "unprecedented." WHO responded by stating that the outbreak was no different from others that had occurred in Africa. Later in 2014, the Director-General of WHO, Margaret Chan, admitted that she was not fully informed of the outbreak's complexity.

WHO finally acknowledged that the Ebola epidemic constituted a global health emergency on August 8, 2014, by declaring a **Public Health Emergency of International Concern (PHEIC)**. The WHO Director-General is the only person who can issue a PHEIC and only after consulting with their executive committee on the proper course of action. There are reasons for the length of time required for its issuance. A PHEIC is

FIGURE 5.1 The world headquarters of the World Health Organization, Geneva, Switzerland.

© Martin Good/Shutterstock

an instrument that asks member nations of the United Nations to participate in disease prevention, surveillance, and control. WHO has only issued this declaration twice before: during the 2009 pandemic of H1N1 (swine flu) and the 2014 reemergence of poliovirus. Certain conditions expressed in WHO's **International Health Regulations (IHR)** must also be met before such a step is taken. The emergency must represent a danger of international spread and require a coordinated international response for its containment. The Director-General must first convene the IHR Emergency Committee, and based upon their recommendation a PHEIC may be issued on a temporary basis.

Border Closures and Flight Restrictions

As an early response, some countries closed their borders to air travel into and out of affected countries in West Africa, plus Chad, Kenya, and South Africa. A number of African countries imposed entry restrictions from Guinea, Liberia, and Sierra Leone in August 2014 that remain in effect. A list of African countries with entry restrictions is presented in TABLE 5.1 . Border closures were a mobility challenge to healthcare workers interested in entering and leaving affected areas, as were air travel border closures that hampered evacuation efforts for infected medical workers. Dr. Anthony Fauci, director of the **National Institute of Allergy and Infectious Diseases (NIAID)**, noted that closing

TABLE 5.1	Entry Restrictions to Africa (as of April 10, 2015)

Flights to Africa

Air France continues suspension of all flights to Sierra Leone since August 2014.

Kenya Airlines continues suspension of all flights to Sierra Leone.

British Airways continues suspension of all flights to Liberia and Sierra Leone until further notice.

Emirate Airlines continues suspension of all flights to Guinea.

Korean Airways has suspended all flights to Kenya.

Note that a traveler should consult the airlines regarding specific flights to the affected countries of Africa before planning a trip as restrictions are in a state of flux.

Countries with Entry Restrictions

Botswana does not allow entry for any noncitizen traveling from Guinea, Sierra Leone, and Liberia until further notice.

Cape Verde bans entry for noncitizens traveling from Guinea, Sierra Leone, and Liberia.

Chad has banned travel from the affected countries (Guinea, Sierra Leone, and Liberia).

Gabon allows entry from the affected countries (Guinea, Sierra Leone, and Liberia) on a case-by-case basis.

Kenya restricts entry from the affected countries except for health professionals attempting to contain the outbreak.

Mauritania has banned entry for all travelers from the affected countries (Guinea, Sierra Leone, and Liberia).

Namibia has banned entry for all travelers from Guinea, Sierra Leone, and Liberia.

Rwanda bans entry for all travelers who have visited the affected countries in 22 days prior to travel.

South Africa has banned entry from the affected countries except for absolutely essential travel considered on a case-by-case basis.

borders and isolating affected countries makes the problem worse by limiting the ability to transport supplies and workers that are essential for containing the epidemic.

In October 2014, the U.S. Department of Homeland Security (DHS) announced that entry into the country for travelers arriving from Ebola-affected areas would be limited to five international airports that already handled 95% of passengers from those countries. Arriving passengers were issued an information kit and a thermometer and requested to monitor their temperature daily. If symptoms such as fever, nausea, vomiting, or diarrhea developed, travelers were asked to report them to state or local health authorities. Failure to comply resulted in DHS officials taking immediate steps to locate the passenger. The U.S. military also implemented its own "controlled monitoring" policy, which meant that personnel returning from Ebola-affected areas would be monitored for 21 days at both domestic and international sites prior to their release into the general population.

While border closures are easier to maintain at airports, the borders between African countries are porous for those traveling on land. In some places, a border consists of little more than a stream or small river that is easily crossed in a small boat. An example of a small body of water constituting a border is shown in **FIGURE 5.2**. Despite this, the country of Guinea closed its border with Sierra Leone in an effort to stamp out the viral disease. Guinea's President Alpha Condé announced in late March 2015 that its borders with Sierra Leone would be closed for 45 days in an effort to eliminate Ebola from the country. The Ebola virus disease continues to infect people in Guinea despite declines in the number of new cases in Sierra Leone and Liberia.

Border closures are having an economic and agricultural effect on the West African countries of Sierra Leone, Guinea, and Liberia. The United Nations reported that border restrictions have impacted the food-processing and production chains in the affected countries, causing a severe shortage of food. The **World Food Programme** (a branch of the United Nations) predicted in December 2014 that by the middle of March 2015 up to 1 million people would be in danger of severe hunger from food shortages. To prevent this crisis, the World Food Programme helped nearly 2.5 million people in the three most affected countries with food assistance. Support included providing survivors

with 3 months of food assistance as well as the purchase of food from local markets and farmers to support the economy. The financial cost of the Ebola crisis from border closures has been severe for the three most affected countries. The World Bank projects that the combined losses will be around 1.6 billion U.S. dollars after calculating the impact on foreign investments, trade, and business activities.

Quarantine and Isolation

Quarantine is a method used to prevent the spread of a disease. It is derived from the Italian words for 40 days, *quaranta giorni*, which was the number of days that the crew of a ship was required to remain aboard before being allowed to disembark during the era of the Black Death (or plague). **Quarantine** is a measure that restricts the movement of people who have been exposed to a contagious disease for observation to determine if they become ill. **Isolation** refers to separating people with a contagious disease from those who are well. The policies regarding quarantine in Liberia and Sierra Leone are summarized in TABLE 5.2.

The enforcement of isolation and quarantines in the United States has been relegated to the Centers for Disease Control and Prevention (CDC). Through its Division of Global Migration and Quarantine, the CDC operates 20 quarantine stations at U.S. ports of entry, which include a number of international airports and a land crossing in El Paso, Texas. Although it has the power to do so, the federal government rarely imposes either isolation or quarantine laws; the last large-scale use of these powers

TABLE 5.2	Policies Regarding Quarantine	
	Liberia	Sierra Leone
Scope of Quarantine	Any person who has had contact with an Ebola virus–infected person	Entire household if any member develops Ebola
Length of Quarantine	21 days post exposure or 2 negative laboratory tests	2 to 21 days post exposure; without symptoms
Ebola—confirmed cases	Immediate isolation	Immediate isolation

was during the 1918–1919 influenza pandemic, which is estimated to have caused the deaths of more than 2 million Americans. A list of contagious diseases that are subject to federal quarantine is shown in **TABLE 5.3** .

In the United States, each state has authority over the quarantine and isolation of its residents. Governors of New York, New Jersey, Illinois, and Florida all ordered quarantines related to healthcare workers who might have had contact with Ebola-infected patients. In New Jersey, a nurse returning from Sierra Leone was placed under a mandatory 21-day quarantine. She was tested on two occasions for the Ebola virus and was found to be negative both times. The White House supported her appeal for release from quarantine.

The controversy generated by the mandatory quarantine mirrors the fear associated with the Ebola epidemic and underscores the public's concern regarding the power of

TABLE 5.3	Diseases for Which Isolation or Quarantine Is Authorized by Executive Order of the President of the United States
Cholera	
Diphtheria	
Tuberculosis (active)	
Plague	
Yellow Fever	
Smallpox	
Viral hemorrhagic fevers (e.g., Marburg virus, Ebola, Lassa fever, Lujo, Crimean-Congo, South American)	
Severe acute respiratory syndrome (SARS)	
Any strain of influenza virus capable of causing a pandemic (such as H1N1 [swine flu])	

both state and federal governments to detain citizens. The best known case in U.S. history of isolation and quarantine is that of Mary Mallon, better known as "Typhoid Mary."

Mary Mallon was an Irish emigrant who worked as a cook for a number of affluent families in the New York City area and was linked to numerous outbreaks of typhoid. She was determined to be an asymptomatic carrier of the disease and is the first person in medical history to receive that designation. She was twice placed in isolation (quarantine) by the New York City Health Department; the second time was for the remainder of her life (23 years).

The laws regulating isolation and quarantine remain basically unchanged since the time of Mary Mallon. Similar decisions have been made in the 21st century regarding HIV-positive individuals who knowingly transmit their illness or tuberculosis patients who carry multidrug-resistant strains of the tubercle bacillus.

▸▸ Safety Measures

The Ebola epidemic in West Africa created a period of heightened awareness in the United States, especially after an Ebola-infected man died in a Dallas hospital. Two nurses caring for the patient also became infected despite reports that they had followed CDC guidelines. The CDC had previously declared that any hospital in the United States could safely care for an Ebola patient. In principle, that statement is true *provided* that healthcare workers are trained in and follow the correct procedures to ensure their safety.

Emphasizing this point, **National Nurses United (NNU)** called for a 2-day strike in November 2014 by its approximately 100,000 members. The nurses protested the lack of a federal directive to hospital administrators on how to protect their workers from Ebola infection. The CDC issued a general set of guidelines from which hospitals could decide on what directives to follow. NNU demanded that its members be issued the best **personal protective equipment (PPE)** that meets the ASTM International (formerly known as the American Society for Testing and Materials) standards F1670 for blood and F1671 for viral penetration. The nurses' union also demanded that air-purifying respirators become part of the PPE. The CDC determined several days after the Dallas nurses became infected that a breakdown occurred in their *removal* of protective equipment.

In October 2014, the CDC issued a comprehensive set of guidelines, which included establishing treatment centers and preparing every hospital in the United States to treat an Ebola patient. Comprehensive information in the form of videos was provided for the correct way to put on and remove PPE. The guidelines stated that before any U.S. health-care worker can become involved with the treatment of an Ebola patient, they must receive comprehensive training and demonstrate competency in infection control. Training, whether for U.S. or international healthcare workers, includes use of a buddy system whereby two people observe each other's actions, scrutinizing for a breakdown in safety when donning or removing PPE.

PPE for most activities includes impermeable gloves (use of double gloves), waterproof boots, face shield or goggles, gown, apron, head covering, and surgical or respiratory mask. Examples of PPE are shown in FIGURES 5.3 through 5.6. Certain procedures, such as autopsies, require additional PPE or the administration of aerosol treatments that mandate a respirator. Fogless goggles are recommended, and goggles may be worn beneath a face shield at the worker's preference. Because heat exhaustion may occur in many environments, breathable gowns may be worn (although they provide less protection) in situations where exposure to infectious agents is low.

FIGURE 5.3 CDC instructors using a mock Ebola treatment center for proper techniques for removing personal protective equipment (PPE).

Courtesy of Cleopatra Adedeji/CDC

FIGURE 5.4 Healthcare workers in Guinea during the 2014 epidemic practicing the proper removal of an Ebola virus–contaminated towel under the supervision of CDC trainers.

Courtesy of Dr. Heidi Soeters/CDC

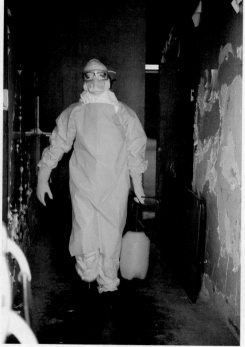

FIGURE 5.5 A participant in the CDC's training course for Ebola management as part of the West African Ebola Response Team. The individual is wearing full personal protective equipment consisting of head covering, goggles, nasal mask, rubber gown, gloves and boots.

Courtesy of Nahid Bhadelia, MD/CDC

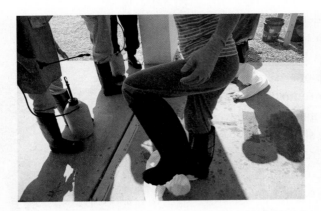

▶▶ Logistics of Healthcare Delivery

The United States took an early leadership role in combatting the Ebola epidemic in West Africa once WHO declared that the epidemic was ongoing. Since March 2014, the United States has sent more than 3,000 federal health officials as part of a 10,000-plus civilian-based response to Sierra Leone, Liberia, and Guinea. The government has also constructed 15 treatment centers, operated 190 burial teams, and provided more than 400 metric tons of supplies including PPE in the affected countries.

In October 2014, President Obama appointed Ron Klain (FIGURE 5.7) to be the "Ebola Czar," with a mission to coordinate federal agencies in their fight against the Ebola virus epidemic. Officially, his title was Ebola Response Coordinator. One of Klain's achievements was obtaining government approval to issue temporary cell phones to all travelers from the affected areas of Africa so that they could be reached for daily monitoring after entering the United States. He was also instrumental in coordinating with the CDC a standardized policy for monitoring all individuals who had traveled to the areas of West Africa affected by the epidemic. On February 15, 2015, Klain returned to the private sector, with President Obama commending his effectiveness in coordinating the U.S. response (FIGURE 5.8).

Operation United Assistance is the operational name for the U.S. response to the Ebola epidemic working through the U.S. Army Africa. To date, approximately 400 soldiers have been deployed to Liberia. Strict protocols have been established to limit any direct personal contact, which amounts to a no-touch and 3-foot separation rule that applies to all local nationals. Twice-daily temperatures are also taken from all troops, and hands are washed in a chlorine solution. Military members are subject to a mandatory 21-day quarantine before their return to the general population. In addition to

■ **FIGURE 5.7** Ron Klain, the Ebola Response Coordinator appointed by President Obama in October, 2014.

Courtesy of Pete Souza/White House

FIGURE 5.8 The President meeting with his cabinet and advisors regarding the Ebola crisis. President Obama cancelled two planned trips in order to remain at The White House and oversee the government's response to Ebola. Meeting with cabinet members and the director of the CDC, the President was reassured by leading public health authorities that the Ebola virus posed no immediate threat to the United States.

Courtesy of Pete Souza/White House

constructing treatment centers, a special 25-bed treatment center was erected in Monrovia, Liberia, By soldiers for the treatment of local healthcare workers.

▸▸ Social Media

Over the past few years, social media has become more than just a way for friends and family to share status updates. Twitter and Facebook provide users with the ability to create content, which now may be considered news. In this way, social media differs from traditional media such as television and radio in which the user is a consumer—rather than a creator—of content. Users of social media typically report news about their personal lives and post photographs with accompanying texts for their friends and relatives to see.

During emergency situations, social media has assumed a completely different function. When Hurricane Sandy devastated the New Jersey coastline in 2012, more than 20 million Twitter posts or "tweets" were sent providing information about the storm. This occurred despite the nearly total loss of functioning cell phones. Following the Boston Marathon bombings, about 56% of 19- to 26-year-olds polled by the Pew Research Center used social media to receive news. The Boston Police Department used social media as a method for disseminating information vital to the eventual capture of the bombing suspects.

A different use for social media was discovered in 2006 by John Brownstein and Clark Freifeld while working at Boston Children's Hospital. Both realized that social media could be used as a means for tracking the spread of infectious diseases. They cofounded **HealthMap** (www.healthmap.org) as a free website. The site provides real-time information on a variety of infectious diseases and utilizes a wide source of information derived from social media comments as well as traditional sources such as WHO. HealthMap uses a **web crawler** to search hundreds of thousands of Internet sources. A dedicated visualization for Ebola outbreaks can be found at www.healthmap.org/ebola.

Near the beginning of the 2014–2015 Ebola outbreak, researchers at HealthMap were able to determine ahead of WHO that the Ebola outbreak was rapidly developing into a multi-country epidemic. On March 14, 2014, HealthMap obtained the following communication by using social media comments published by Africaguinee.com:

> A new disease that we do not know the name was reported in the prefecture of Macenta located 800KM from Conakry, killing 8 people dead...
> Symptoms manifested by "anal and nasal bleeding"
> Seems to resemble Lassa fever.

Eight days later, WHO declared that an epidemic of Ebola was occurring.

▶▶ Fast-Tracking Drugs and Vaccines

Antiviral Agents

When the Ebola outbreak developed into a multi-country epidemic, only two pharmaceutical companies, Mapp Biopharmaceutical and Tekmira Pharmaceuticals, were in the process of developing effective antiviral agents. Mapp Biopharmaceutical is developing **ZMapp**, and Tekmira Pharmaceuticals is developing **TKM-Ebola**, both of which are antiviral drugs; however, they have a different mechanism of action.

ZMapp is a combination of three types of **monoclonal antibodies** that are directed at different stages in the Ebola virus lifecycle. The antibodies are produced by a process called **biopharming** that uses biotechnology to add the genes for antibody production into the chromosomes of plants. In the case of ZMapp, the monoclonal antibodies are extracted from tobacco plants.

Experimental drugs undergo a series of clinical trials designated as Phase 1 through Phase 4 (see TABLE 5.4). ZMapp received approval from the U.S. Food and Drug Administration (FDA) on February 21, 2015, to begin Phase 1 clinical trials for the treatment of Ebola virus disease. Phase 1 is a small study using 20 to 80 healthy volunteers in an effort to determine safety and an effective dose. The drug was used to treat seven patients who had been diagnosed with Ebola infection. Two of the seven receiving ZMapp died.

TKM-Ebola is classified as a **small interfering RNA (siRNA)** drug. These are very small molecules that interfere with the expression of specific genes. TKM-Ebola blocks the formation of three of the seven proteins that are part of the complete Ebola virus particle. In January 2014, TKM-Ebola began Phase 1 of its clinical trials. The trial was put on a hold basis after some recipients developed upper respiratory tract infections similar to influenza. During the Ebola crisis, the FDA added the drug to its **expanded access** category, which means that the drug can be given for treatment of patients with confirmed infection with Ebola virus but not to other categories of patients.

TABLE 5.4	Clinical Trial Phases	
Phase	Participants	Goals
1	20–80 healthy volunteers	Determine dosage guidelines, document how a drug is metabolized and excreted, and identify acute side effects
2	100–300 people with the disease or condition that the product could possibly treat	Gather further safety data and preliminary evidence of the drug's beneficial effects and develop and refine research methods for future trials with the drug
3	1,000–3,000 people with the disease or condition that the product could possibly treat	Further test the product's effectiveness, monitor side effects, and compare the product's effects to a standard treatment, if one exists
4	Varied. Conducted after a product is already approved and on the market	Learn more about the treatment's long-term risks, benefits, and optimal use, or to test the product in different populations of people, such as children

Source: Data from U.S. Food and Drug Administration. (2014). *Inside clinical trials: Testing medical products in people*. Retrieved from http://www.fda.gov/Drugs/ResourcesForYou/Consumers/ucm143531.htm

Vaccines

Two types of vaccines are used today based on the form of virus used: inactivated and live attenuated. An **inactivated vaccine** has been treated with heat of chemicals so that it cannot cause disease in a person following vaccination. The poliovirus vaccine used today in the United States is an example of an inactivated vaccine. A **live attenuated vaccine** contains elements of the replicating virus but has been treated so that it is less virulent. An example of live attenuated vaccines is the measles, mumps, and rubella (MMR) vaccine that is given to infants as part of their vaccine regimen.

Viral diseases are best prevented rather than treated. For this reason, a vaccine to prevent the Ebola virus is the best method for preventing a recurrence of the Ebola epidemic of 2014. To this end, four pharmaceutical companies are developing vaccines that are undergoing clinical evaluations to determine their safety and effectiveness. Presently, no vaccine exists that has received approval for routine vaccination.

A fast-track program has been developed for the Ebola vaccine. An international consortium consisting of the Center for Infectious Disease and Policy located at the University of Minnesota (in the United Sates) and the Wellcome Trust (in the United Kingdom) have convened a group of 26 experts to develop a "roadmap" for fast-tracking the vaccine's development.

© Vladnik/Shutterstock

The consortium released its report, *Development of Ebola Vaccines: Principles and Target Product Criteria,* in January 2015. It emphasized the need for vaccines, because the epidemiology of the Ebola virus is likely to change during the 21st century, causing more outbreaks as well as the possibility of a **pandemic** or worldwide epidemic. The report is viewed as a "living document," meaning it is subject to revision as more is learned about the Ebola virus, and it may help to expedite vaccine testing and development. The need for global community involvement is emphasized and considers all aspects of vaccine production, pricing, and distribution to the most affected West African countries.

GlaxoSmithKline and NIAID are testing a vaccine identified as ChAd3-ZEBOV based on an inactivated adenovirus, while the Public Health Agency of Canada and Merck Vaccines are testing a vaccine based on vesicular stomatitis virus identified as rVSV-ZEBOV. "ZEBOV" stands for Zaire Ebola virus; the Zaire strain is the one causing the outbreak in West Africa. All vaccines under study are inactivated and contain parts of the Ebola virus but are not live attenuated versions. For this reason, the vaccines are considered completely safe with no possibility of trial volunteers developing Ebola virus disease.

Johnson & Johnson, in collaboration with Bavarian Nordic, is developing a two-dose vaccination that uses different vaccines for the first and second doses. Novavax is also developing a recombinant vaccine based on proteins from the Guinea 2014 strain of Ebola virus.

Phase 1 trials for the ZEBOV vaccines are currently in progress and Phase 2 and 3 trials are scheduled to begin in 2015. Phase 1 trials for the Johnson & Johnson vaccine will begin in early 2015. The availability of the vaccines will depend upon outcomes of the clinical trials. As of this writing, only the two ZEBOV vaccines are under clinical efficacy testing and a licensed vaccine is anticipated by the end of 2015.

CRITICAL THINKING QUESTIONS

1. Why do you think that the world's leading healthcare organizations were slow to respond to the Ebola outbreak in the early months of the epidemic?

2. If you were a politician, would you support closing your country's borders to travel from countries that were part of the Ebola outbreak? Would your answer be different for residents and nonresidents?

3. Why isn't there more international investment in an Ebola vaccine?

4. Make the case in support of the use of quarantine on the grounds of public health.

5. Make the case against quarantine on the grounds of individual liberty.

6. How do you think the case of Typhoid Mary would be treated today?

The Future of Ebola and Other Emerging Diseases

Carolyn Dehlinger

Should we worry about the future of Ebola? In the grand scheme of human health and disease, how big of a threat is Ebola? Looking at the numbers, deaths from the Ebola virus pale in comparison to malaria or human immunodeficiency virus (HIV). Malaria resulted in 660,000 deaths in 2010, while HIV caused the deaths of 1.5 million people in 2013. Ebola is not as contagious as tuberculosis or cholera, which can spread through the air and water, respectively. Tuberculosis is one of the world's deadliest diseases, with about one-third of the entire human population infected. Cholera kills approximately 100,000 people each year. The H1N1 virus (swine flu) killed hundreds of thousands during a pandemic in 2009–2010. As of late March 2015, the current Ebola outbreak breeched the 10,000 mark of recorded deaths. Is it rational to focus such enormous resources on a pathogen with comparatively low impact upon human health?

Regardless of any comparison to other communicable diseases, Ebola remains an example of the overall preparedness of worldwide health agencies in responding to severe outbreaks of any pathogen—or, lack of preparedness, as it were. The H1N1 pandemic that began in 2009 served as a case study in how unprepared the world was for such an event. In 2010 the World Health Organization (WHO) convened its Internal Health Regulations (IHR) review committee to analyze the response to the pandemic and concluded, "the world is ill-prepared to respond to a severe influenza pandemic or to any similarly global and threatening public-health emergency." In general, WHO and other human health agencies are better equipped to handle short-term, localized response efforts and less able to deal with long-term, global coordination of a sustained campaign against a particular disease. The world tends to react better to acute, immediate emergencies as compared to chronic and prolonged threats. At the time, the IHR made terse recommendations on how to improve the overall response capabilities of WHO and its allies, most of which were ignored. A 2014 report by the IHR noted that just 64 of 194 member states of WHO met the 2010 recommendations, which included sufficient laboratory infrastructure, data management, and surveillance of outbreak progression. This became evident during the most recent Ebola outbreaks. Thus, the future of Ebola is linked to how we are able to react to all disease outbreaks and, furthermore, how we might prevent their occurrence altogether.

■ QUESTIONS TO CONSIDER

Some questions to consider as you read this chapter include:

1. How might we better monitor the introduction and spread of communicable diseases such as Ebola?

2. How can we trace the inevitable evolution of Ebola into different strains and potentially different modes of transmission?

3. What other emerging diseases should we be concerned about and why?

4. What plans are being adopted to control this disease and prevent further outbreaks?

Room for Improvement: Shortcomings of the 2014–2015 Outbreak Response

The unfortunate truth of the Ebola outbreak in West Africa that began in 2014 is that the healthcare systems of the most affected countries are woefully unprepared, under-staffed, and underfunded. As treatment facilities became overwhelmed with patients, Ebola became the focus for healthcare workers, while other patients and diseases had resources diverted. In April 2015 WHO released a statement outlining what it perceives to be the greatest shortcomings of the 2014–2015 international response. The organization readily admits that gains in other human health concerns, such as lower rates of infant mortality, inroads against malaria, and more women surviving childbirth, are built upon a vulnerable foundation. WHO noted directly that healthcare systems in areas affected by Ebola are fragile in this way and further declared that both national and international systems as they exist today are inadequate to properly handle widespread epidemics of this scale.

Because of the great scare the Ebola virus caused, virtually every aspect of life in the affected countries was impacted. Schools and markets were shut down, local borders were closed, and the disruption to daily lives resulted in chaos and the spread of misinformation. Rumors, like Ebola being airborne, abounded. Normal business activities ceased and caused stultifying effects on the local economies. Developing countries like Guinea, Liberia, and Sierra Leone already struggle with economic and environmental concerns, which further contribute to the ability of emerging diseases like Ebola to take hold. Outbreaks only worsen these conditions. This creates a negative feedback loop—countries with poor healthcare systems granting limited access to its citizens end up confounding the effects of a highly fatal outbreak as they struggle to confine and treat patients.

There is also limited ability to educate its populace. This in turn perpetuates the spread of disease that might otherwise be prevented through cautionary measures, if only the knowledge were sufficiently communicated (FIGURE 6.1). A combination of factors may have contributed to this situation, such as language and cultural barriers, economic limitations, and scarcity of resources such as infrastructure and technology. WHO noted in its assessment of the 2014–2015 international response that there was a general lack of effective communication. WHO highlighted the need to better address cultural differences by more effectively framing the messages they deliver to the affected regions. The organization also recommended being willing to receive information from local people, including the full consideration of cultural practices. It is equally important that local leaders be transparent in their reporting of emergence and incidence so that response efforts are based on the best possible information.

Countries that are able to detect and confirm Ebola cases and then provide the best available treatment methods to patients provide a stark contrast to those just described. For example, Nigeria, Mali, and Senegal all experienced confirmed cases of Ebola within

FIGURE 6.1 A classroom in Guinea where the CDC's International Infection Control Team relays information on the science behind Ebola hemorrhagic fever. The training is expected to reach at least 2,000 healthcare workers in the most affected prefectures of Guinea: Nzérékoré, Macenta, and Kérouané.

Courtesy of Dr. Heidi Soeters/CDC

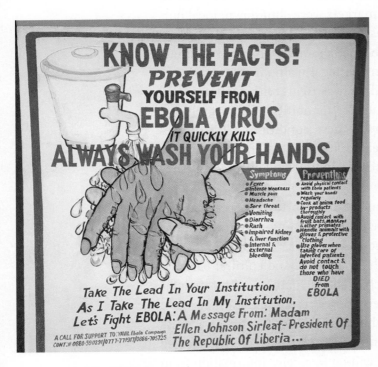

FIGURE 6.2 A public service poster seen in 2014 in Liberia advising on the importance of hand washing for preventing the spread of Ebola. This poster was created for the President of the Republic of Liberia, Ellen Johnson Sirleaf.

Courtesy of Sally Ezra/CDC

their boundaries. However, these countries were able to minimize the spread of the virus due to better preparedness. These countries had sufficient clinical laboratory support necessary to identify and isolate infected patients. Because they had significant healthcare infrastructure already in place, these countries were capable of responding much more quickly and providing the necessary treatment to their patients. Having these capabilities effectively prevented the spread of Ebola before it became a major epidemic in these countries, unlike in places such as Sierra Leone or Guinea.

Another lesson of the 2014–2015 Ebola outbreaks is the need for careful coordination of all preventive and response measures. Public education campaigns are important (**FIGURE 6.2**). Rapid diagnostic testing in order to identify new cases is necessary. But these alone will not suffice in stamping out the spread of Ebola. Patients must remain isolated until test results are confirmed. Treatment units must effectively administer the best available care, contact tracing must continue, and contaminated

FIGURE 6.3 This image shows a makeshift incinerator made from a used oil drum for use in safe disposal of contaminated items and hazardous wastes. Seen in Lagos, Nigeria.

Courtesy of Daniel DeNoon/CDC

waste must be properly disposed of in order to prevent release into the environment (FIGURE 6.3). Accordingly, community response to the Ebola outbreak may indeed require a change in existing cultural practices such as burial traditions that involve direct contact with the deceased. We cannot expect organizations like the Centers for Disease Control and Prevention (CDC), WHO, and Doctors Without Borders to work within the confines of existing beliefs and cultural practices. Modifying burial practices will be more effective than reacting with sweeping condemnation and refusal to respect tradition. The community itself, not just medical professionals and organizations, must participate in prevention efforts. Inroads that involve community leaders such as tribal chiefs and religious leaders should be expanded. All interested parties must cooperate or we will continue to witness the lingering effects of this outbreak. Therefore, in order to effectively engage the general populace, there needs to be a concerted effort to provide incentives toward their participation. Incentives would include being respectful of centuries-old practices and seeking the advice and opinions of local people while still making clear any modifications are in the best interest of everyone. Of course, survival would be the greatest incentive of all.

In addition to this, for the best possible implementation and practice of any and all measures, the biggest obstacles we still face are transportation and communication. With limited connectivity between towns and villages, poor road conditions, and unreliable transport vehicles, it is difficult to get infected patients to appropriate treatment facilities. It also increases the time it takes to send off specimens and receive corresponding test results. Many of the affected areas have little to no Internet access, and other telecommunications have unreliable coverage or are not present. This has prevented effective communication of patient data and other statistics. For successful outbreak containment, information needs to be conveyed and disseminated in a more centralized manner. As the global response to Ebola continues, we must address these concerns if we are to improve conditions and ultimately see the end of this outbreak.

▸▸ Ethical Considerations

There has been an ongoing conversation during this Ebola outbreak regarding the ethics of care and response. Perhaps most notably, there is a discrepancy in the care provided to local populations of affected countries as compared to what is given to Western medical workers who have contracted the virus. While indigenous patients remain within their country, where many receive relatively rudimentary treatment, some foreign aid workers have been transported to world class facilities in Europe and the United States. This has resulted in a dichotomy between recovery rates, with a 71% fatality rate in local patients compared to a 26% fatality rate of infected Western patient cases. The concern about disparity in healthcare services may arise as we see experimental drugs and vaccines being tested in affected countries. Will there be delays in providing the best possible care due to concerns over potential profits? Though the reason for their decision was not explicitly provided, we have already seen at least one pharmaceutical company (Chimerix) pull out of clinical trials due to declining incidence of Ebola cases in Liberia. Some wonder if the focus on treatments and vaccines is too shortsighted and reactionary. WHO recommends the goal of creating an environment geared toward the long-term, broader picture. To do so, a standardized, worldwide system of response to outbreaks and epidemics should include some consideration of the ethical implications involved.

▸▸ Spillover: Species Jumping

Viruses, while not technically alive, are capable of evolving into new strains with distinct characteristics. For example, the concept of **spillover** describes how viruses can gain entry into new species they had previously been unable to infect. This is made possible by changes to the virus anatomy that make it capable of infecting a new type of cell. It may become compatible with a receptor protein on a new type of host cell, allowing it to gain entry when in the past it could not. When a disease initially takes hold in one species and then jumps to another, it is referred to as a spillover event. The virologist Nathan Wolfe (2014) wrote in the *Wall Street Journal* of two emerging viruses worth noting in addition to the Ebola virus: These viruses result in Middle East respiratory syndrome (MERS) and avian influenza (H7N9). Both emerged concurrently with the

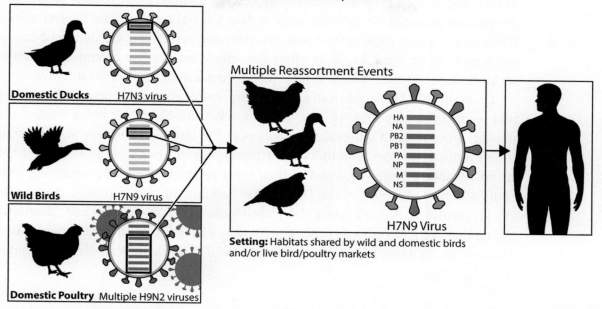

Genetic Evolution of H7N9 Virus in China, 2013

Domestic Ducks H7N3 virus

Wild Birds H7N9 virus

Domestic Poultry Multiple H9N2 viruses

Multiple Reassortment Events

HA
NA
PB2
PB1
PA
NP
M
NS

H7N9 Virus

Setting: Habitats shared by wild and domestic birds and/or live bird/poultry markets

The eight genes of the H7N9 virus are closely related to avian influenza viruses found in domestic ducks, wild birds and domestic poultry in Asia. The virus likely emerged from "reassortment," a process in which two or more influenza viruses co-infect a single host and exchange genes. This can result in the creation of a new influenza virus. Experts think multiple reassortment events led to the creation of the H7N9 virus. These events may have occurred in habitats shared by wild and domestic birds and/or in live bird/poultry markets, where different species of birds are bought and sold for food. As the above diagram shows, the H7N9 virus likely obtained its HA (hemagglutinin) gene from domestic ducks, its NA (neuraminidase) gene from wild birds, and its six remaining genes from multiple related H9N2 influenza viruses in domestic poultry.

Centers for Disease Control and Prevention
National Center for Immunization and Respiratory Diseases

FIGURE 6.4 The process of viral reassortment resulted in the new strain of H7N9 avian influenza.

Courtesy of Dan Higgins (Illustrator)/CDC

most recent Ebola outbreak, and both represent spillover events. WHO's Global Alert Response website (http://www.who.int/csr/outbreaknetwork/en/) reports outbreaks significant to human health, and this list is populated with MERS and strains of avian influenza. H7N9 provides an example of the process that makes jumping species possible for a virus. Known as **reassortment**, a new viral strain can emerge when two or more different viruses infect the same host cell (FIGURE 6.4). While occupying the same cell, the viruses are able to exchange genes responsible for coat receptor proteins. This results in new combinations of receptors and, along with them, a new viral strain. Close contact among viral host species further enables reassortment, because it increases the likelihood of more than one virus infecting a single host cell.

Dr. Wolfe (2014) promotes the establishment of a concerted and sustained effort in monitoring what he calls "viral chatter." Instead of waiting for an outbreak to occur, a network of existing and constantly staffed monitoring stations equipped with rapid diagnostic testing laboratories should be in place in order to identify emerging diseases in real time. Such a network should ideally extend beyond Africa to other locales with the greatest potential for spillover, such as Asia, Central America, and South America. When people are in close contact with reservoir and vector species, spillover becomes much more likely. Thus, in places where live animal markets are common, or hunting for bushmeat is a typical source of food, monitoring should be the most intensive.

Dr. Wolfe (2014) also advocates a global preventive strategy as opposed to our current tendency to play what he calls "pandemic Whac-A-Mole." He notes the fact that smaller, more isolated outbreaks of Ebola over the last 40 years were warning signs of an eventual widespread outbreak. Furthermore, countries with a history of Ebola outbreaks appear better equipped in their most recent response. For example, when Ebola was first observed in the Democratic Republic of Congo (DRC), the country quickly identified its index case and deployed mobile laboratories for onsite diagnostic testing that reduced response times and increased isolation of infected patients. All of this contributed to containment of the virus in DRC, while in West Africa the spread of Ebola continues.

A Model Country

About 75% of new viruses are zoonotic in origin. In order to truly prevent human diseases before they reach epidemic, outbreak, or even pandemic proportions, monitoring of animal and human populations is most desirable. Cameroon is a model country in the spillover-monitoring initiative. The government of this country has

© Nata-Lia/Shutterstock

instituted two major programs designed to identify emerging diseases and limit their spread. The National Program for Zoonosis Prevention and Control and the National Emergency Committee for Pandemics are involved in educating healthcare workers and the general public about pathogens of zoonotic origin, establishing diagnostic and treatment centers, and coordinating research efforts designed to identify and monitor diseases in animal and human populations. However, developing countries where this type of monitoring is needed cannot act alone. Future efforts will require financial and technical assistance from the developed world along with occasional deployment of healthcare teams when outbreaks do occur.

PREDICT

PREDICT is a global monitoring surveillance organization aimed at identifying spillover events and predicting the spread of **emerging infectious diseases**. The organization is a partnership among the U.S. Agency for International Development (USAID), the University of California at Davis School of Veterinary Medicine, the Wildlife Conservation Society, the EcoHealth Alliance, Metabiota, and the Smithsonian Institution. Metabiota is a company founded by Nathan Wolfe. PREDICT was instrumental in identifying the introduction of Ebola into DRC in 2014.

The Global Animal Information System (GAINS) is utilized by PREDICT as a data-collection and dissemination tool. Essentially, it is a database used to compile the extensive amount of information obtained in the PREDICT system. About 20 countries are involved in the project. As of this writing, scientists working with PREDICT have collected around 200,000 tissue specimens from approximately 45,000 animals in the search for zoonotic organisms. GAINS enables researchers to compile data on these specimens into a searchable and remotely accessible form. In this manner, PREDICT can use GAINS as a tool for modeling the emergence of spillover events, effectively identifying and predicting their behavior. Indeed, PREDICT is responsible for the identification of more than 800 viruses previously unknown to us. GAINS is managed by Metabiota and also makes use of data reported from external sources. As a computer technology platform, GAINS is capable of compiling information from 50,000 websites each hour and thus collects reports on zoonotic threats as they emerge.

In addition to managing GAINS and assisting with the global efforts of PREDICT, Metabiota acts as a consulting agency to the governments of developing countries. The company is able to assist in risk assessments of different geographic areas, make recommendations on monitoring projects, and provide expertise in data analysis and spillover emergence. The scientists of Metabiota are also actively collecting samples and detecting novel viral species. They engage in education of government agencies and communities at large, working to implement effective prevention and other control strategies. This is all taking place in an age where global monitoring is more important than ever. Global travel networks allow viruses and other pathogens to spread amongst populations like never before (FIGURE 6.5). Additionally, climate change results in shifting habitats, where reservoir and vector species spread into new territories, bringing zoonotic pathogens with them.

FIGURE 6.5 A video display presentation was created by the CDC and posted in the Hartsfield-Jackson Atlanta International Airport during the 2014 West African Ebola outbreaks.

Courtesy of Daniel DeNoon/CDC

▸▸ Emerging Zoonotic Diseases

Zoonotic diseases are a major concern for human health because of their severe and often fatal consequences. Because they are new to our species, typically this means we have no innate defense against them. We lack antibodies that would enable our immune systems to fight them off, because we have no previous exposure to their antigens. In addition to the *Filoviridae* family, of which the Ebola virus is a member, several viruses causing viral hemorrhagic fevers (VHFs) can be considered emerging diseases with similar potential to wreak havoc on human populations. The most recent Ebola outbreak has heightened the concerns over VHFs in general and serves as a lesson in how to better prevent the spread of these deadly viruses. The *Arenaviridae, Bunyaviridae*, and *Flaviviridae* families all cause VHFs worth considering as present and future threats. Likewise, the *Henipavirus* genus of the *Paramyxovirus* family of viruses is currently involved in spillover to the human population. As noted earlier, MERS and avian influenza are also concerning. In this section, a brief overview of each family and the specific diseases they cause is presented.

Arenaviridae

The *Arenaviridae* family of viruses, of which some are hemorrhagic, are transmitted to humans by rodents. Usually, a person will pick up one of the viruses through breathing in aerosolized urine or dust particles containing fecal material. These sources may also enter through mucous membranes like the eyes, nose, or mouth, as well as by gaining entry through breaks in dermal tissue. At least five viruses from this family are known to cause human disease (TABLE 6.1). Person-to-person contact as a method of transmission is well documented with Lassa and Lujo fevers.

Lassa Fever

Lassa fever is named for the village in Nigeria where it was first identified. In 1969, it killed its first two known victims, both missionary nurses. Caused by the Lassa virus shown in FIGURE 6.6 , this disease is transmitted by its reservoir host, the multimammate rat (*Mastomys natalensis*), which is indigenous to western Africa. Many of the same countries experiencing the most recent Ebola outbreaks are likewise endemic to Lassa fever. Approximately 100,000 to 300,000 people become infected each year, with around

TABLE 6.1	Arenaviruses and Associated Diseases
Arenavirus	**Human Disease**
Chapare virus	Chapare hemorrhagic fever
Guanarito virus	Venezuelan hemorrhagic fever
Lymphocytic choriomeningitis virus (LCMV)	Lymphocytic choriomeningitis
Junin virus	Argentine hemorrhagic fever
Lassa virus	Lassa fever
Machupo virus	Bolivian hemorrhagic fever
Sabiá virus	Brazilian hemorrhagic fever

Source: Data from Centers for Disease Control and Prevention. (2013). Viral Hemorrhagic Fevers. Retrieved from http://www.cdc.gov/ncidod/dvrd/spb/mnpages/dispages/vhf.htm

5,000 deaths. The overall fatality rate is 15–20% of all patients hospitalized for Lassa fever, with much higher rates in pregnant women, where 95% of fetuses die in utero.

Lujo Hemorrhagic Fever

The Lujo virus, discovered in 2008, causes Lujo hemorrhagic fever (LUHF). To date, LUHF is rare but highly fatal, killing around 80% of those infected. After Lassa fever was observed in 1969, Lujo fever was the second VHF from the *Arenaviridae* family to occur in Africa. It is more common in southern Africa, with the first case occurring in Zambia. Field workers, along with their cohabitants and sexual partners, are the most susceptible to Lujo virus infection. As with the Lassa virus, Lujo virus is transmitted from a rodent reservoir, with secondary person-to-person contact resulting in new cases as well. Nosocomial infections are a notable concern with LUHF.

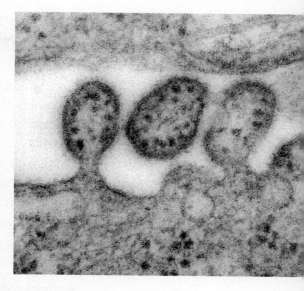

FIGURE 6.6 A transmission electron micrograph of three Lassa virions. The center virion has just budded from the host cell shown at the bottom of the image. The two on each side are in the process of budding.

Courtesy of C. S. Goldsmith/CDC

Bunyaviridae

The *Bunyaviridae* family of vector-borne viruses typically infects insects, rodents, and occasionally plants. Certain strains are capable of infecting humans, with insects or rodents acting as vectors. These viruses are similar to Ebola and other filoviruses with regard to their anatomy and genetic makeup. They are also enveloped and contain single-stranded, negative-sense RNA as their genetic material. *Bunyaviridae* contain five genera of viruses classified as either Biosafety Level 3 or 4 by the CDC ( **TABLE 6.2**).

TABLE 6.2	*Bunyaviridae* Genera and Associated Diseases
Genus	**Associated Disease Example(s)**
Hantavirus	Hemorrhagic fever with renal syndrome (HFRS) Hantavirus pulmonary syndrome (HPS)
Nairovirus	Crimean-Congo hemorrhagic fever (CCHF)
Orthobunyavirus	La Crosse encephalitis
Phlebovirus	Rift Valley fever
Tospovirus	Affects plants only

Source: Data from Centers for Disease Control and Prevention. (2013). Viral Hemorrhagic Fevers. Retrieved from http://www.cdc.gov/ncidod/dvrd/spb/mnpages/dispages/vhf.htm

Crimean-Congo Hemorrhagic Fever (CCHF)

Transmitted to humans via tick bites, certain birds, or ruminants, Crimean-Congo hemorrhagic fever (CCHF) has a 10–40% fatality rate with death occurring around 2 weeks after infection. Individuals who survive infection typically improve 9 to 10 days after contracting the disease. The *Hyalomma* genus of ticks is the primary vector of the disease, as well as its reservoir (FIGURE 6.7 , inset). These ticks exhibit a geographic range throughout the Balkans, Africa, the Middle East, and Asia, south of the 50° parallel, as shown in Figure 6.7. Humans may also become infected with CCHF from infected ruminants, such as cattle, during their slaughter or while handling raw meat. In fact, most recorded human cases of CCHF have occurred in people who work with livestock or in slaughterhouses. Additionally, human-to-human transmission is

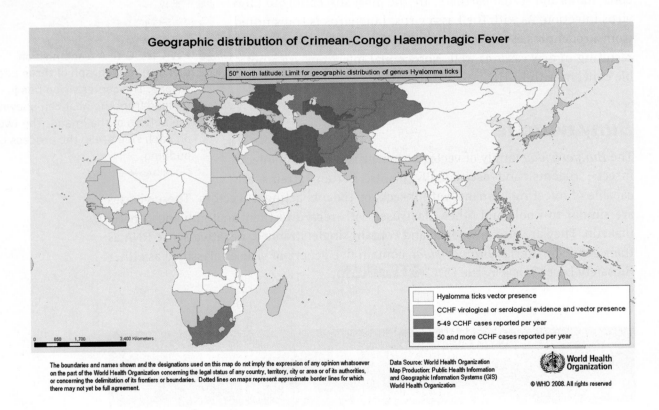

Geographic distribution of Crimean-Congo Haemorrhagic Fever

50° North latitude: Limit for geographic distribution of genus Hyalomma ticks

	Hyalomma ticks vector presence
	CCHF virological or serological evidence and vector presence
	5-49 CCHF cases reported per year
	50 and more CCHF cases reported per year

0   850   1,700   3,400 Kilometers

The boundaries and names shown and the designations used on this map do not imply the expression of any opinion whatsoever on the part of the World Health Organization concerning the legal status of any country, territory, city or area or of its authorities, or concerning the delimitation of its frontiers or boundaries. Dotted lines on maps represent approximate border lines for which there may not yet be full agreement.

Data Source: World Health Organization
Map Production: Public Health Information and Geographic Information Systems (GIS) World Health Organization

World Health Organization

FIGURE 6.7 A map of the Eastern hemisphere shows the geographic distribution of Crimean-Congo hemorrhagic fever. Ticks of the *Hyalomma* genus (inset) are the vector species that transmits Crimean-Congo Hemorrhagic Fever to humans. The top image shows a dorsal view; the bottom shows a ventral view.

Reprinted from World Health Organization, 2008, www.who.int. Inset: © Armando Frazao/Shutterstock

Crimean-Congo Hemorrhagic Fever (CCHF) Virus Ecology

Enzootic Cycle
Ixodid (hard) ticks are both a reservoir and vector for the CCHF virus.

The virus is maintained in nature transovarially and transstadially.

Epizootic-Epidemic Cycle
CCHF cases occur more during the warmer parts of the year, mostly the spring and summer. There are no cases during the winter.

Humans become infected through tick bites and direct contact with infected animal blood or tissue.

Ticks feed on numerous wild and domestic animals such as cattle, goats, sheep, birds, and hares. These animals serve as both food sources for ticks and amplifying hosts for the CCHF virus.

Transmission can occur while slaughtering infected animals, during veterinary procedures, and in hospital settings where proper protective equipment and appropriate disinfection procedures are lacking.

Adult

Eggs

Nymph

Larva

CDC

FIGURE 6.8 The ecology of Crimean-Congo hemorrhagic fever shows the enzootic and epizootic cycles for the virus.
Courtesy of CDC

possible in a manner similar to transmission of Ebola virus (through contact with bodily fluids of an infected individual). Nosocomial infections are therefore possible as well. An overview of the CCHF lifecycle is shown in FIGURE 6.8. Because there is no vaccine available for CCHF, there is concern for a potential epidemic. Currently, we must rely on physical preventive measures, such as avoiding tick bites and exercising caution when handling livestock.

Rift Valley Fever

First observed in eastern Africa, Rift Valley fever (RVF) was responsible for an outbreak among sheep in 1930. This virus is transmitted by mosquitoes and is endemic to most parts of sub-Saharan Africa, with cases also occurring in the Arabian Peninsula (FIGURE 6.9). In 2000, cases were reported in both Saudi Arabia and Yemen, with the most recent outbreak occurring in Kenya during 2006–2007. RVF is considered an **epizootic** disease, meaning it primarily infects animals (FIGURE 6.10). Epizootic diseases are a concern for humans when we come into close contact with infected animals. Because RVF is well documented in livestock populations, humans are likewise at risk due to our interdependent relationship with these animals (FIGURE 6.11). For example, in 1977 more than 600 people succumbed to RVF in Egypt, where the infection was ascribed to livestock imported from the Sudan. In spite of these large

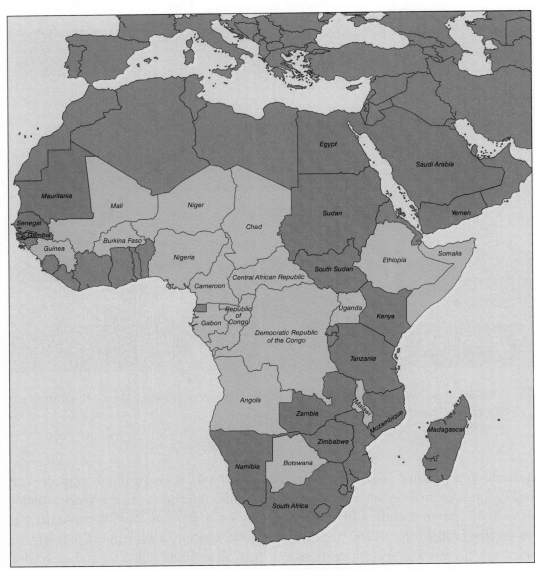

RIFT VALLEY FEVER DISTRIBUTION MAP

- Countries reporting endemic disease and substantial outbreaks of RVF
- Countries reporting few cases, periodic isolation of virus, or serologic evidence of RVF infection
- RVF status unknown

0   250   500   1,000
Miles

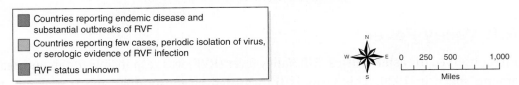

FIGURE 6.9 The geographic distribution of Rift Valley fever.

Courtesy of CDC

numbers of fatalities, in general RVF is not fatal, and most infected people recover within 2 days to 1 week of symptom onset. However, approximately 1% of infected individuals will develop hemorrhagic fever, and among these the fatality rate is 50%. Additionally, around 1% of patients will develop **encephalitis** (swelling of the brain), which may be fatal.

Rift Valley Fever (RVF) virus ecology

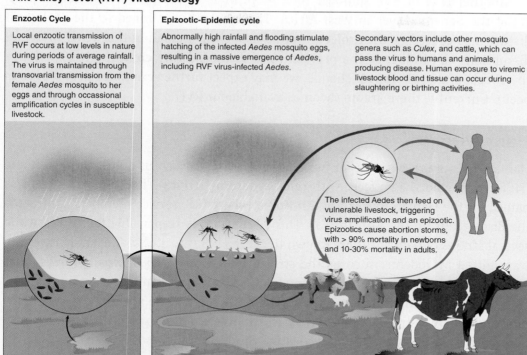

Enzootic Cycle

Local enzootic transmission of RVF occurs at low levels in nature during periods of average rainfall. The virus is maintained through transovarial transmission from the female *Aedes* mosquito to her eggs and through occassional amplification cycles in susceptible livestock.

Epizootic-Epidemic cycle

Abnormally high rainfall and flooding stimulate hatching of the infected *Aedes* mosquito eggs, resulting in a massive emergence of *Aedes*, including RVF virus-infected *Aedes*.

Secondary vectors include other mosquito genera such as *Culex*, and cattle, which can pass the virus to humans and animals, producing disease. Human exposure to viremic livestock blood and tissue can occur during slaughtering or birthing activities.

The infected Aedes then feed on vulnerable livestock, triggering virus amplification and an epizootic. Epizootics cause abortion storms, with > 90% mortality in newborns and 10-30% mortality in adults.

FIGURE 6.10 The ecology of Rift Valley fever shows enzootic and epizootic cycles. Mosquitoes are both the reservoir and vector of this disease.

Courtesy of CDC

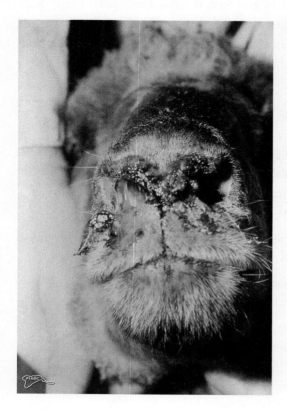

FIGURE 6.11 This image depicts an ovine sick from Rift Valley fever in Mexico. Taken by the Mexico-United States Commission for the Prevention of Foot and Mouth Disease for the 1982 publication, "Illustrated Manual for the Recognition and Diagnosis of Certain Animal Diseases."

Courtesy of Dr. Jerry J. Callis, PIADC/Dr. Brian W.J. Mahy/CDC

Another case of RVF spillover took place in 1987 during a construction project at the Senegal River in West Africa. The incident was linked to the ecological disturbance caused by the project, serving as a clear indication that environmental disruptions can influence the prevalence of disease. As we continue to experience the effects of global climate change, we may expect further examples of spillover to occur. Currently, there are no vaccines available for RVF.

Hantaviruses

Hantaviruses are responsible for two types of disease: hantavirus pulmonary syndrome (HPS) and hemorrhagic fever with renal syndrome (HFRS). The Hantaan virus was named for the Hantan River in South Korea, where U.S. service members contracted the disease during the Korean War. This viral strain is responsible for HFRS, while several strains of hantavirus are known to cause HPS (TABLE 6.3). Hantaviruses are transmitted by rodents, such as the striped field mouse (*Apodemus agrarius*) that serves as the vector for HFRS. The mode of transmission is much the same as with arenaviruses. Rodents that carry the viral strains responsible for HPS are very common and widely distributed throughout North America. For example, the deer mouse (*Peromyscus maniculatus*) that carries the Sin Nombre virus (SNV) strain responsible for most cases of HPS in the United States is present in most states, Canada, and Mexico (FIGURE 6.12).

Person-to-person transmission is extremely rare and unreported in the United States. Fatality due to HFRS is between 5 and 15% of infected patients. While HPS fatality is on average 36%, rates have ranged between 19 and 56% (FIGURE 6.13a). Thirty-four states in the United States have reported cases of HPS, most notably in states west of the Mississippi River (FIGURE 6.13b). During the most recent outbreak in 2012, three out of 10 persons infected while visiting the Yosemite National Park did not survive.

TABLE 6.3	Rodent Vectors of New World Hantaviruses	
New World Hantavirus Strain	**Vector Species of Rodent**	**Habitat**
Bayou Virus (BAYV)	Marsh rice rat (*Oryzomys palustris*)	Marshlands
Black Creek Canal Virus (BCCV)	Cotton rat (*Sigmodon hispidus*)	Grasses and shrubs
New York Virus (NYV)	White-footed mouse (*Peromyscus leucopus*)	Woodlands, brush, open areas
Sin Nombre Virus (SNV)	Deer mouse (*Peromyscus maniculatus*)	Woodlands and deserts

Source: Data from Centers for Disease Control and Prevention. (2012). Rodents in the United States That Carry Hantavirus. Retrieved from http://www.cdc.gov/hantavirus/rodents/index.html

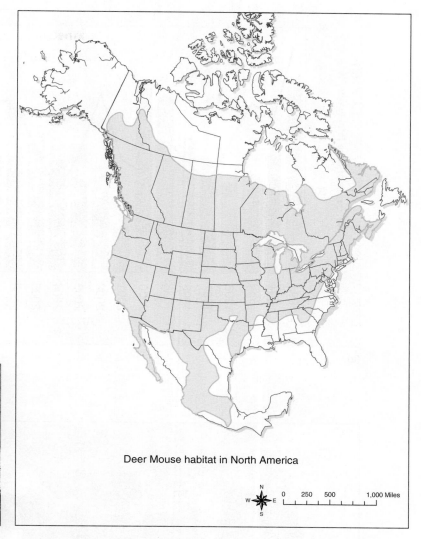

Deer Mouse habitat in North America

FIGURE 6.12 The deer mouse *Peromyscus maniculatus* (inset) that carries the Sin Nombre virus strain responsible for most cases of hantavirus pulmonary syndrome in the United States is present in most states, Canada, and Mexico.

Courtesy of CDC
(inset) Courtesy of James Gathany/CDC

There are currently no treatments or vaccines for hantaviruses. The primary preventive measure for avoiding hantavirus infection is limiting exposure to rodents in addition to practicing good housekeeping.

Flaviviridae

The *Flaviviridae* are a family of single-stranded RNA viruses. Ticks and mosquitoes are two known arthropod vectors that transmit flaviviruses to humans. Viruses transmitted through arthropods are known as **arboviruses**. Most notable as emerging infectious diseases are West Nile virus, yellow fever, and dengue fever.

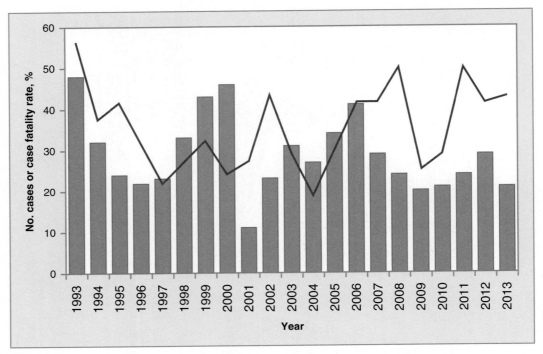

(a)

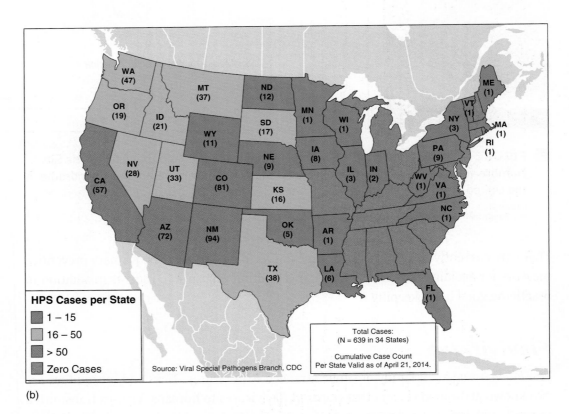

(b)

FIGURE 6.13 (a) Annual cases of hantavirus pulmonary fever and fatality rates in the United States, 1993–2013. (b) Cases of hantavirus pulmonary fever by state of reporting.

Courtesy of CDC

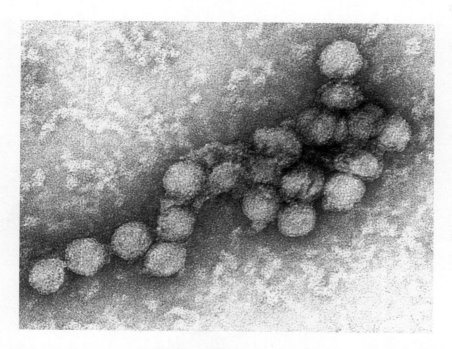

■ **FIGURE 6.14** A colorized transmission electron micrograph image of the West Nile virus.

Courtesy of Cynthia Goldsmith/CDC

West Nile Virus

Serving as a clear example of spillover, West Nile virus (WNV) infected and killed North American bird populations before entering the human species. More than 300 species of birds have tested positive for WNV; however, the disease is most fatal to species of crows and jays. When WNV was first discovered in the United States in 1999, it was a common practice to seek public reports of suspicious dead birds. As of now, WNV has been detected in all 48 continental states, and monitoring of dead birds has fallen to the wayside. Depicted in **FIGURE 6.14**, WNV is the most commonly contracted arbovirus in the United States. Because WNV is transmitted through mosquito bites, incidence tends to increase in the summer months. Insect repellants and protective clothing are the primary means by which individuals can avoid exposure, while municipalities combat WNV transmission through destruction of mosquito breeding areas and application of insecticides to standing water. The CDC partners with the U.S. Geological Survey in the reporting system ArboNET, which monitors the incidence of all arboviruses within the United States.

Dengue Fever

Humans have known of dengue fever since it was first reported in 1789. At that time, it was known as "breakbone fever," because it causes severe pain in the joints and muscles. It is endemic in Puerto Rico and many tropical islands within the Caribbean, as well as in Central and South America, Southeast Asia, and the Pacific island chains (**FIGURE 6.15**). Like WNV, it is transmitted through mosquito vectors of the *Aedes*

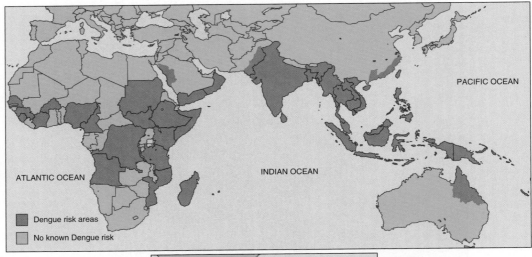

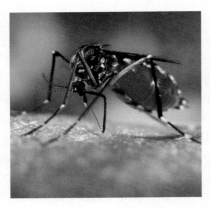

FIGURE 6.15 This map shows the geographic distribution of Dengue virus, with more than 100 countries endemic to the pathogen. A female mosquito of the *Aedes aegypti* species (inset) acts as a vector for Dengue viruses.

Inset: Courtesy of James Gathany/CDC. Map: Courtesy of CDC

genus (Figure 6.15 inset). Four dengue viruses (dengue 1–4) are associated with the human diseases dengue fever (DF), dengue shock syndrome (DSS), and dengue hemorrhagic fever (DHF). Up to 400 million people are infected with one or another of the dengue viruses each year.

It is known that dengue virus originated in a nonhuman primate reservoir somewhere between 100 and 800 years ago, making it a relatively old "emerging" disease. Again, human activity seems to have played a role in its increased incidence within the last century. Until the First World War, dengue-associated illnesses were quite rare. With increased displacement of individuals due to both world wars, along with the increased transportation of cargo vessels, came the unintentional transport of mosquito vectors into new regions of the world. At present, about 22,000 deaths result each year from dengue viral infections.

Yellow Fever

Yellow fever is caused by the yellow fever virus (FIGURE 6.16 , inset), named such due to the oft-associated jaundice seen in infected patients. The yellow fever virus has three modes of entry into the human species. The sylvatic (or jungle) cycle requires transmission from nonhuman primates in forested areas through mosquito vectors into humans. A savannah (or grasslands) cycle is common to geographical areas that border jungles and essentially is an offshoot of the sylvatic cycle. Here, mosquitoes may pass yellow fever back and forth between nonhuman primates and humans or from human to human, such as individuals living near forest borders. Lastly, the urban cycle passes yellow fever from forest or savannah-inhabiting humans to city-dwelling mosquitoes, which then infect other humans (Figure 6.16). This last cycle is how the disease is spread in highly populated areas.

Yellow fever is endemic to sub-Saharan Africa and South America. Up to 30,000 deaths result from yellow fever each year, with more than 90% of those reported in Africa. Increased incidence since the mid-1990s is attributed to a combination of climate change, deforestation, and human migration patterns such as displacement and urbanization. Thankfully,

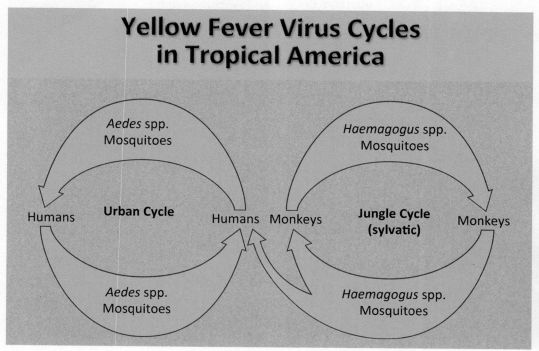
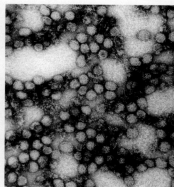

FIGURE 6.16 The "Yellow Fever Virus Cycles in Tropical America" produced by the CDC shows the sylvatic and urban lifecycles. The inset shows a photomicrograph image of yellow fever virions magnified 234,000 times.

Courtesy of CDC
Inset: Courtesy of Erskine Palmer, PhD/CDC

with yellow fever there is a vaccine. This can be used in combination with mosquito control measures in order to prevent spread of the disease. WHO has implemented a vaccination program in cooperation with the United Nations Children's Fund (UNICEF) and local governments within Africa. To date, 12 countries have participated in the program.

Paramyxoviridae

In 1999, encephalitis and respiratory illness broke out among adult men residing in Malaysia and Singapore. Upon investigation, a previously unknown virus was determined as the cause. Named Nipah virus, it was later classified into the *Paramyxoviridae* family of viruses (FIGURE 6.17). Pigs act as the reservoir host for Nipah virus, with transmission to humans occurring when in close proximity to this livestock animal. Nipah virus has a fatality rate of up to 80%, placing it among the deadliest emerging diseases discussed in this chapter. The reservoir for Nipah virus is unknown, but preliminary studies suggest bats known as flying foxes, such as those of the genus *Pteropus*, are the most likely candidates (FIGURE 6.18).

Nipah virus belongs to the genus *Henipavirus* (HNV) along with its close relative, Hendra virus. They are the only paramyxoviruses classified as Biosafety Level 4 by the CDC. HNV is distributed throughout Southeast Asia, Australia, and India. Outbreaks have coincided with significant levels of deforestation within these regions. This is

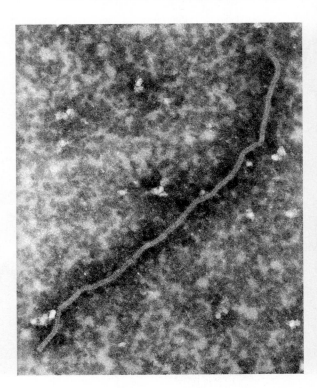

FIGURE 6.17 A transmission electron micrograph of the Nipah virus.

Courtesy of C. S. Goldsmith; P. Rota/CDC

FIGURE 6.18 The spectacled flying fox (*Pteropus conspicillatus*) is one species of fruit bat suspected as a reservoir for Nipah virus.

Courtesy of Brian W.J. Mahy, BSc, MA, PhD, ScD, DSc/CDC

another instance of environmental disturbance, whether anthropogenic or otherwise, resulting in emerging disease incidence. Recent studies by PREDICT team members have found evidence to suggest spillover may be in the near future within African countries, as *Pteropodidae* bat species are routinely hunted there for bushmeat.

Middle East Respiratory Syndrome

MERS is caused by a coronavirus (Middle East respiratory syndrome coronavirus, or MERS-CoV) and was first reported in 2012 (FIGURE 6.19). An index case in Saudi Arabia has led to subsequent spread of the virus throughout the world, including the United States. The CDC has the authority to detain individuals traveling into the United States who are suspected of carrying the virus. Coronaviruses also include the virus responsible for severe acute respiratory syndrome, or SARS. The animal reservoir is not definitively known, but MERS-CoV has been found in camels. Therefore, a working hypothesis suggests humans may be infected through close contact with these animals. This emerging infectious disease is so new that scientists are actively investigating the virus. There is still much to learn about MERS-CoV.

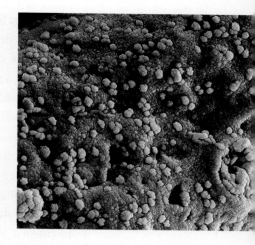

FIGURE 6.19 A scanning electron micrograph image of MERS-CoV virions (shown in yellow) infecting a host cell (shown in blue) was produced by the National Institute of Allergy and Infectious Diseases (NIAID).

Courtesy of National Institute of Allergy and Infectious Diseases (NIAID)/CDC

Avian Influenza

In April 2013, WHO reported a newly discovered strain of H7N9 avian influenza capable of infecting humans (FIGURE 6.20a). Reassortment among other H7N9 strains likely

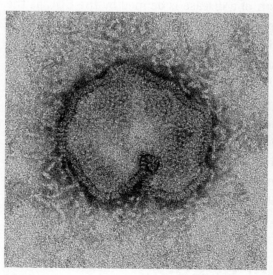

(a)

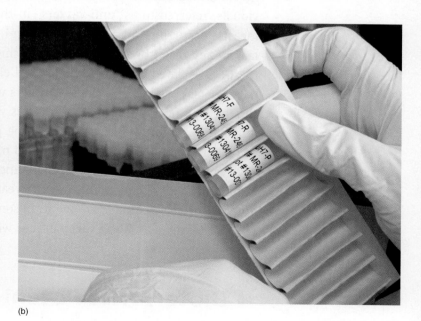

(b)

FIGURE 6.20 (a) A scanning electron micrograph image of the new strain of H7N9 virions, an avian influenza virus. (b) A laboratory worker holding reagents from a rapid diagnostic kit for the new strain of H7N9 avian influenza, developed by the CDC.

(a) Courtesy of Cynthia S. Goldsmith and Thomas Rowe/CDC. (b) Courtesy of Rob Taylor/CDC resulted in this new strain. This strain causes severe respiratory illness, which sometimes results in death. Approximately one-third of cases are fatal. To date, with one exception, cases have been isolated to China, with all patients having made contact with infected poultry. A sole case reported in Malaysia in February 2014 was an individual who had recently traveled to affected areas within China. The CDC has developed a rapid diagnostic kit designed for reverse transcriptase polymerase chain reaction (RT-PCR) detection of the strain (**FIGURE 6.20b**). Health officials are most concerned about H7N9 and other influenza strains, because they reassort so readily. There is significant potential for a strain to spill over or become more infectious or virulent. Concerns that the virus may spread to neighboring bird and/or human populations have resulted in the close monitoring of this and other influenza strains. They are among the most likely emerging infectious diseases to cause a global pandemic.

▶▶ The Role of the Anti-Vaccination Movement

A growing movement in the Western world has people consciously choosing not to vaccinate their children for a variety of reasons. Misinformation has led to confusion over the safety and efficacy of vaccines. Parent groups and blogs exacerbate the spread of rumors and fear. For example, a common misbelief is that the rise in cases of autism in the United Staes is the result of vaccines tainted with mercury. Some of these individuals, referred to as "anti-vaxxers" in the media, are linked to a recent outbreak of measles occurring in California. **Elimination** of a disease is defined by the CDC as the lack of any new transmission events within a period of 12 or more months. Although the CDC declared the United States had eliminated measles as of 2000, sporadic cases have popped up from individual travelers who bring the disease back home with them. In late 2014, a new outbreak of measles was announced. Stemming from an infected individual visiting the Disneyland theme park, 121 individuals representing 17 states and the District of Columbia contracted measles. Of those cases, 103 are confirmed to have visited Disneyland within a small range of dates. However, the index case has yet to be identified. As scientists race toward creation of an Ebola vaccine, it is worth considering this anti-vaccination movement. In some future scenario, could we see the approval and marketing of an Ebola vaccine faced with widespread refusal of its administration?

▶▶ Bioterrorist Potential of Ebola

In February 2014, the United States launched the Global Health Security Agenda. It is a pledge to aid 30 countries in monitoring infectious disease and other biological threats, including those with bioterrorism potential. Bioterrorism is the use of microorganisms to introduce harm to a population, whether by directly infecting humans or tainting water and/or food supplies. Because of its high fatality rate, Nipah virus has been targeted by researchers as a potential bioterrorism agent. Nevertheless the CDC lists Nipah virus as a Category C bioterrorism agent. This is the lowest threat rating, based upon availability, the ease with which it may be synthesized for mass dissemination, and the morbidity and mortality of the infectious agent. Much higher on the list, Ebola is classified by the CDC as a Category A bioterrorism agent. Category A includes the deadliest of diseases such as plague, anthrax, smallpox, and botulism (see TABLE 6.4). Along with other VHFs, Ebola is ranked as a high-priority threat because of its high mortality rates. VHFs may be misdiagnosed as other disease with similar symptoms, such as the flu. That could lead to a delay in response to an act of bioterrorism. The presentation of VHFs further compounds their potential effectiveness as a bioterrorism agent, along with other important characteristics. The disease carries a stigma, as we have seen with the reactions among the general public demonstrated through social and other media outlets. This stigma can increase the potential for overreaction, fear, and pandemonium in the public sphere. Bioterrorism would likely rely upon the general social disruption and chaos as much as, if not more than, the actual transmission of disease.

TABLE 6.4	**Category A Agents and Diseases They Cause**				
Disease	**Agent(s)**	**Method of Transmission**	**Incubation Time***	**Symptoms**	**Treatment**
Anthrax	*Bacillus anthracis*	Inhalation	1–7 days	Flu-like symptoms, fatigue, breathing difficulty, possible death	Antibiotics (early administration), prevention by vaccination
Botulism	*Clostridium botulinum* toxoid	Foodborne	12–36 hours	Double vision, blurred vision, slurred speech, difficulty swallowing, dry mouth, muscle weakness, paralysis of breathing muscles leading to death	Antibiotics
Plague (pneumonic)	*Yersinia pestis*	Inhalation	1–6 days	Fever, headache, cough, weakness	Antibiotics
Smallpox	Variola virus	Inhalation	7–17 days	Characteristic rash, fever, fatigue	Prevention by vaccination (within 4 days after exposure)

(continues)

TABLE 6.4	Category A Agents and Diseases They Cause (*Continued*)				
Disease	Agent(s)	Method of Transmission	Incubation Time*	Symptoms	Treatment
Tularemia	*Francisella tularensis*	Inhalation	1–14 days	Fever, ulcerated skin lesions, chills and shaking, headache, weakness; can progress to respiratory fever and shock	Antibiotics
Viral hemorrhagic fevers	Ebola virus, Marburg virus, Lassa virus	Close contact with infected person's body fluids	7–14 days (Ebola fever), 5–10 days (Marburg fever), 6–21 days (Lassa fever)	Fever, fatigue, dizziness, weakness, internal bleeding, bleeding from body orifices, shock, delirium	Supportive therapy, ribavirin (depending on circumstances)

*Considerable variation

Source: Data from Centers for Disease Control and Prevention. (2015). Emergency Preparedness and Response: General fact sheets on specific bioterrorism agents. Retrieved from http://emergency.cdc.gov/bioterrorism/factsheets.asp

▶▶ Sexually Transmitted Infection

Sexually transmitted infections (STIs) are pathogens passed from person to person during sexual relations via the exchange of bodily fluids. These may be seminal fluid, vaginal secretions, saliva, and/or blood. At present, Ebola is not a designated STI. However, WHO recommends caution due to scattered evidence that suggests the virus may persist in the aforementioned fluids. To date, investigations into the potential STI status of Ebola have been limited. This is an area in which we would expect to see further study, especially as the duration of the outbreak progresses and recovering patients return to their homes and lives. What scarce information is available has primarily been obtained as footnotes during studies with other goals. A comprehensive and systematic investigation into Ebola's potential as an STI is still an unexplored but desirable avenue.

The nature of science can be thought of as fumbling in the dark to some extent, where scientists do not always know what they are looking for until they find it. Exploring one area of interest may lead to something entirely unexpected, yielding that *"Eureka"* moment. Along with this notion, scientists may make assumptions about the nature of something in order to err on the side of caution. In this case, we follow the "better safe than sorry" approach when it comes to whether Ebola is a viable STI. Researchers have detected traces of Ebola viruses in the seminal fluid of convalescent male patients for up to 3 months (at least 82 days) following recovery from the disease. Therefore, WHO advises any recovering Ebola patient to abstain from any type of sexual contact for at least 3 months. In lieu of this recommendation should it not be possible, the organization advises the use of male and/or female condoms in order to prevent the potential spread of viral particles. As for female patients, there is less evidence supporting the persistence of virions in vaginal secretions. A lone sample taken from a female patient 33 days following her recovery resulted in the detection of trace particles of the virus, but researchers did not confirm whether live virus was present.

As of late March 2015, the CDC has reevaluated its transmission guidelines to include contact with semen of male survivors as "possibly" spreading the virus. Like WHO, the CDC continues to recommend avoiding sexual contact for three months for all survivors following recovery from the disease, whether male or female, until more information is known. Additionally, the CDC now recommends all male survivors properly and consistently use condoms during any form of sexual contact, regardless of the time elapsed since recovery. Guidelines on safe disposal of used condoms and subsequent thorough washing of skin with soap and water are also included.

The new recommendations were in response to the report of a possible sexual transmission from a male survivor to a female in Monrovia, Liberia. The infected female tested positive for Ebola using RT-PCR diagnostic methods. Contact tracing revealed the patient had no means of contracting Ebola other than vaginal intercourse with a male survivor. The male survivor was discharged from a Liberian ETU in October 2014, free from RT-PCR detection in his blood and lacking any physical symptoms of Ebola. Contact with the infected female did not occur until five months later, exceeding existing recommendations of a three-month waiting period. Upon confirmation of the female infection, the male survivor underwent RT-PCR and ELISA of his blood and semen. The blood remained free of Ebola but did indicate Ebola antibodies. RT-PCR testing of his semen revealed a partial detection of Ebola RNA with a 28% sequence match to the female's infection. In spite of this new development, the CDC remains cautious in defining Ebola as an STI, stating it may "possibly" be spread through contact with semen from a male survivor. It is still not definitively proven whether the male survivor infected the female patient. Further genomic testing and live cultures will determine if his semen did contain viable infectious Ebola particles and whether there is a complete genetic match to the virions detected in the infected female. Further studies are expected, as they will be necessary to confirm the potential nature of Ebola as an STI.

▸▸ Hindsight Is 20/20: Prevention of Ebola

The future of the Ebola virus will greatly depend on our ability to prevent its spread into not-yet affected countries. This will rely on the coordination of various countries to monitor their shared borders and prevent the passage of infected individuals into different locations. We will need to improve the accuracy and availability of contact lists. Ideally, these lists could be shared with border checkpoints and airports so that any contact could be identified on site. There is also potential for the development of portable and rapid diagnostic tests. We could systematically test individuals and vet them before allowing passage.

Improvements to existing healthcare infrastructure will also aid in preventing the spread of Ebola virus disease and other emerging infectious diseases. As the most recent outbreaks have shown, current facilities were unable to absorb the rapid influx of infected patients, and widespread collapse of healthcare systems followed. This resulted in an environment of fear and confusion that only compounded the problem. Many locals avoided treatment even with the onset of symptoms over concern they would not receive proper care and/or worries over being forced into isolation. Part of the improvements needed in the healthcare system rely on educating the public. People need to have confidence in their healthcare system if they are to trust it and use it. An extensive information dissemination campaign is needed that will improve the public's perception of the system.

An outbreak of Ebola occurring in the DRC in 1995 spurred the development of prevention guidelines. Theoretically, if implemented 20 years ago, we may have prevented the current outbreak. Jointly developed by the DRC Ministry of Health, WHO, and the CDC, the guidelines were specific to VHFs in rural Africa. One of the greatest contributions leading to the outbreak in 1995 was improper identification. Medical professionals in DRC failed to correctly diagnose the first patient, and the incidence of Ebola was not recognized for several months. This span of time allowed the chain of transmission to branch, out and the outbreak peaked before it was even noticed. Thus, there was a primary recommendation in the guidelines to improve diagnostic techniques and the general ability of medical professionals to recognize VHFs. Additionally, the guidelines outlined the need for proper PPE and preventive behaviors that limit the spread of disease in the form of nosocomial cases. Widespread education of in-field medical workers is imperative to this goal. Overall, the preventive guidelines developed in 1995 required healthcare facilities in rural Africa to take a precautionary role rather than a reactionary one. The guidelines took into account the limited resources common to healthcare systems in developing countries such as DRC. Thus, low-cost

FIGURE 6.21 A photograph of CDC team lead, Dr. Ellen Dotson instructing on how to properly wash hands with hyperchlorinated water disinfected with bleach. A concerted effort to place hand-washing stations such as this one throughout West Africa was coordinated by instructing Guinean healthcare workers in an attempt to curb the spread of the Ebola. Taken in Guinea during the 2014 West African Ebola outbreaks.

Courtesy of Dr. Heidi Soeters/CDC

practices were provided, such as the use of household bleach in disinfecting surfaces and equipment (FIGURE 6.21). Despite the efforts in producing these preventive guidelines, they were not fully implemented nor were they distributed effectively to areas with high risk of VHF incidence. It does raise the question: Are we destined to repeat the past? Can we learn from this kind of mistake, or will human nature condemn us to future outbreaks? Only time will tell.

CRITICAL THINKING QUESTIONS

1. What are the ethical concerns surrounding the Ebola outbreak response?

2. How can we balance the need for research and development of treatments, vaccines, and response efforts with the profit-driven nature of any businesses that become involved?

3. Is there potential for future Ebola vaccines to be rejected by the anti-vaccination movement?

4. How could Ebola vaccines be effectively marketed in order to gain widespread adoption?

5. What issues might be raised in the event that systematic rapid diagnostic testing becomes enacted, such as at airports or international borders?

Glossary

Analytical epidemiology: A field of study that applies statistical inferences about health-related states or events in the population based on sample data.

Antibiotic resistance: The ability of a bacterial species to develop resistance to the effect of an antibiotic.

Antibody: A highly specific protein produced by the body in response to a foreign substance such as a bacterium or virus that is capable of binding to the substance.

Antigen: A chemical substance that stimulates the production of antibodies by the immune system.

Antigenic drift: A minor variation over time in the antigenic composition of a virus.

Antisense RNA: The sequence of RNA complementary to sense mRNA; used to block translation.

Arbovirus: A virus transmitted by arthropods (e.g., insects).

Biopharming: The synthesis or production of pharmaceuticals via genetic engineering.

Biotechnology: The commercial application of genetic engineering using living organisms.

Bushmeat: Meat from nondomesticated animals such as bats and monkeys that may be eaten in rural parts of Africa and Asia.

Capsid: The protein coat that encloses the genome of a virus.

Case definition: A term used by epidemiologists to establish a set of unified conditions for an illness that is used by public health officials for surveillance.

Centers for Disease Control and Prevention (CDC): A federal agency under the U.S. Department of Health and Human Services responsible for monitoring the state of public health within the United States.

Chain of infection: A method of analyzing an illness in terms of six links that form a chain of transmission.

Chimeric antibodies: Monoclonal antibodies containing portions from more than one species.

Contact history: Identification of an infected person with follow-up contacting of each person with whom the infected person has had contact.

Contact tracing: A method used by epidemiologists to trace every contact that an infected patient has had with another person with the intent of determining the health status of every contact.

Descriptive epidemiology: A method of epidemiological investigation that attempts to determine the source of an outbreak via the collection of data regarding person, place, and time.

Diagnostic test: A test used in medicine to identify or confirm the presence of a disease or determine disease severity.

Double-blind: A situation where neither the participants nor the assessing investigator(s) know who is receiving the active treatment.

Elimination: The point at which there have been no new transmissions of a disease within a period of 12 or more months.

Emerging infectious disease: An infectious disease that is completely new to medical science or a recognized infectious disease that has increased in number of cases.

Encephalitis: Inflammation of the brain.

Endemic: A disease that is always present within a population at a low rate of incidence and prevalence throughout the calendar year.

Endocytosis: A process of engulfment of material, including viruses, displayed by some cells.

Endosomal sorting complexes required for transport (ESCRT): Protein complexes that are involved in any activity that requires remodeling of the cell membrane.

Envelope: The flexible membrane of protein and lipid that surrounds many types of viruses.

Enzyme-linked immunosorbent assay (ELISA): A serological test in which an enzyme system is used to detect an individual's exposure to a pathogen.

Epidemic: An increase in the number of cases of an illness above the expected number within a population that is widespread.

Epidemic curve: A visual display of the accepted date of onset of an illness, showing an interval of time in days or months.

Epidemiology: A branch of medical science that studies diseases within a population in terms of their causes, distribution, and methods of control.

Epizootic: Describes an outbreak of disease that primarily infects animals.

Expanded access: The approved use of an experimental drug by patients who do not meet the criteria for a clinical trial.

Experimental epidemiology: The use of mathematical models to predict the rate of transmission and the areas into which an infectious disease might spread.

Genetic engineering: The use of bacterial and microbial genetics to isolate, manipulate, recombine, and express genes.

Genetically modified organism (GMO): An organism whose DNA has been artificially altered.

Genome: The complete genetic information of an individual.

HealthMap: An organization that provides real-time information on a variety of emerging public health concerns from sources ranging from social media comments to the World Health Organization (www.healthmap.org).

Hemorrhagic fever: Any of a group of severe, distinct viral illnesses characterized by fever and damage to the vascular system.

Hill's criteria of causation: A set of minimal conditions necessary to establish a causal relationship between two items.

Hybridoma: The cell line resulting from the fusion of a B cell and a cancer cell. Hybridomas are used to produce antibody molecules that will react with only one substance and no other.

Inactivated vaccine: A vaccine that has been created by treating the virus with chemicals or heat to render it noninfectious.

Incidence: The number of newly diagnosed cases of a disease within a population.

Incubation period: The time from exposure to an infectious agent until the development of symptoms.

Index case: The first case of a particular disease or condition to come to the attention of public health authorities.

International Health Regulations (IHR): A set of legally binding regulations from the World Health Organization to help member countries prevent and handle health risks with international potential.

Isolation: The separation of persons who have a specific infectious illness from those who are healthy and the restriction of their movement to stop the spread of that illness.

Koch's postulates: Four steps that are used to verify that an infectious agent is the cause of an illness. The postulates are limited to bacterial diseases and have been modified to include other infectious agents.

Live attenuated vaccine: A vaccine that contains a weakened but still viable version of a virus.

Margin of error: A confidence level based on statistical analysis that a result is true.

Mode(s) of transmission: The route used by a microorganism for spreading within a community. Modes of transmission include water, aerosol droplets, food, and insects among others.

Molecular cloning: A process that utilizes recombinant DNA technology to "clone" myriad copies of a molecule.

Monoclonal antibodies (MAbs): A type of antibody created from a single cellular line, designed to bind to only one substance.

Mutation: A permanent alteration of a DNA sequence.

National Institute of Allergy and Infectious Disease (NIAID): A branch of the National Institutes of Health that conducts and supports research on infectious, immunologic, and allergic diseases.

National Institutes of Health (NIH): A biomedical research facility that is part of the U.S. Department of Health and Human Services.

National Nurses United (NNU): The largest professional association and union of registered nurses in the United States.

Negative-sense: Describes viral RNA that is complementary to the viral mRNA and, therefore, must be changed into positive-sense RNA before it can begin to translate itself into proteins within the host cell. The negative-sense genome then becomes viral messenger RNA (mRNA).

Nosocomial transmission: Spread of a disease or infection in a medical care setting.

Nucleocapsid: The combination of genome and capsid of a virus.

Obligate parasites: An organism that is dependent on a host cell for replication.

Operation United Assistance: The United States' response to the 2014–2015 Ebola epidemic, primarily working through the U.S. Army Africa.

Outbreak: A localized epidemic.

Pandemic: A worldwide epidemic.

Pathogenic: Capable of causing illness in an infected host.

Person under investigation (PUI): A person who has risk factors for or displays signs and symptoms consistent with a disease.

Personal protective equipment (PPE): Equipment worn to minimize exposure to and protect the wearer from infection, illness, or injury.

Placebo: An inactive substance or treatment given to satisfy a patient's expectation of treatment or as a control variable as part of a clinical trial.

Pleomorphism: Variability in the size, shape, and appearance of a microbe.

Point source outbreak: An outbreak that begins with exposure to a common agent such as contaminated food.

Positive-sense: Describes viral RNA that, upon entering the host cell, can begin to translate into proteins; it is of the same form as mRNA.

Prevalence: The number of newly diagnosed and uninfected members of a population.

Prodrug: A precursor of a drug that must undergo a chemical conversion before becoming active.

Propagated outbreak: An outbreak that is transmitted by person-to-person contact, with multiple incubation periods.

Public Health Emergency of International Concern (PHEIC): A formal recognition by the World Health Organization of an international public health risk that may require immediate, coordinated, international action.

Quarantine: The separation and restriction of the movement of persons who, although not yet ill, are suspected of exposure to an infectious agent and may become infectious themselves.

Rate of infectivity: The speed with which a microorganism can infect members of a population.

Reassortment: The recombination of genetic material that often occurs when two or more viruses infect the same cell.

Recombinant DNA technology: The process of combining plasmid transformation with restriction enzymes to produce a unique technology involving the recombination of DNA sequences, thus creating a transgenic organism.

Reproduction number: The number of people that are infected by exposure to an infected person.

Reservoir: The location or organism where disease-causing agents exist and maintain their ability for infection.

Rivers' Postulates: A modification of Koch's postulates that includes viruses as the cause of an infectious disease.

RNA interference: A gene control mechanism that uses RNA as a silencing molecule that interferes with normal gene expression.

SEIR model: A model similar the SI and SIR models, but considers those who have become exposed (E) to a disease who have not yet developed symptoms.

Sense mRNA (RNAi): The coding sequence of mRNA.

Sexually transmitted infection (STI): An infection that is normally passed from one person to another through sexual contact.

SI model: A method determining the spread of a disease based on the number of susceptible (S) and infected (I) individuals.

SIR model: A model that predicts the number of infected members in a closed population based on the number that are susceptible, can become infected (I), and are removed (R) from the population either by recovery or death.

Sign: An objective finding related to an illness, such as temperature or pulse.

Small interfering RNA (siRNA): A class of double-stranded RNA molecules that disrupt the expression of specific genes.

Spillover: Movement of a pathogen from one species to another.

Symptom: A patient's subjective description of how they feel, such as, "I have a headache."

TKM-Ebola: An experimental siRNA therapy designed to target the Ebola virus.

Transgene: Foreign genetic material that is introduced to an organism.

Vectors: An organism that transmits the agents of disease from one host to another.

Viral hemorrhagic fever (VHF): A severe syndrome of the body affecting multiple organ systems with multiple symptoms, including signs of bleeding, whether beneath the skin, within internal organs, or from body orifices like the mouth, nostrils, ears, or anus.

Virion: A completely assembled virus outside its host cell.

Web crawler: An Internet program that systematically browses the World Wide Web, primarily for indexing.

World Health Organization (WHO): An agency of the United Nations that was established in 1948 and is charged with monitoring global public health issues.

World Food Programme: A branch of the United Nations that is the world's largest humanitarian, food-assistance agency.

Zaire ebolavirus: One of five species of Ebola virus—an RNA virus that causes a complex set of symptoms including in its most extreme form massive internal and external hemorrhaging.

ZMapp: An experimental biopharmaceutical drug comprised of a combination of antibodies directed at different stages in the Ebola virus lifecycle.

Zoonosis: An infectious agent that normally resides within an animal population and can be transmitted to humans.

Zoonotic disease: A disease spread from another animal to humans.

Zoonotic spillover: The movement of a pathogen from an animal reservoir into humans.

References

▶▶ Chapter 1

Alexander, K. A., Sanderson, C. E., Marathe, M., Lewis, B. L., Rivers, C. M., Shaman, J., … Eubank, S. (2015). What factors might have led to the emergence of Ebola in West Africa? *Public Library of Science*. Retrieved from http://blogs.plos.org/speakingofmedicine/2014/11/11/factors-might-led-emergence-ebola-west-africa/

Asokan, G. V., & Kasimanickam, R. K. (2013). Emerging infectious diseases, antibiotic resistance and Millennium Development Goals: Resolving the challenges through One Health. *Central Asian Journal of Global Health, 2*(2).

Calisher, C. H. (2014). Lifting the impenetrable veil: From yellow fever to Ebola hemorrhagic fever and SARS. *Journal of Emerging Infectious Diseases, 20*(3).

Centers for Disease Control and Prevention. (2012). *Principles of epidemiology in public health practice: An introduction to applied epidemiology and biostatistics* (3rd ed.). Retrieved from http://www.cdc.gov/ophss/csels/dsepd/SS1978/SS1978.pdf

Centers for Disease Control and Prevention. (2015). *Outbreaks chronology: Ebola virus disease*. Retrieved from http://www.cdc.gov/vhf/ebola/outbreaks/history/chronology.html

Gonzalez, J. P., Pourrut, X., & Leroy, E. (2007). Ebolaviruses and filoviruses. *Current Topics Microbiology Immunology, 315*, 363–387.

Marí Saéz, A., Weiss, S., Nowak, K., Lapeyre, V., Zimmermann, F., Düx, A., … Leendertz, F. H. (2014), Investigating the zoonotic origin of the West African *Ebola* epidemic. *EMBO Molecular Medicine, 7*(1), 17–23.

United Nations Development Programme. (2014). *Human development report 2014: Sustaining human progress: Reducing vulnerabilities and building resilience*. Retrieved from http://hdr.undp.org/sites/default/files/hdr14-report-en-1.pdf

von Drehle, David (2014, December). Person of the Year: The Ebola Fighters. *Time*. Retrieved from http://time.com/time-person-of-the-year-ebola-fighters

World Health Organization. (2015). *Ebola*. Retrieved from http://apps.who.int/ebola/

▶▶ Chapter 2

Belyi, V. A., Levine, A. J., & Skalka, A. M. (2010). Unexpected inheritance: Multiple integrations of ancient Bornavirus and Ebolavirus/Marburgvirus sequences in vertebrate genomes. *PLoS Pathogens, 6*(7), e1001030.

Bray, M., & Geisbert, T. W. (2005). Ebola virus: The role of macrophages and dendritic cells in the pathogenesis of Ebola hemorrhagic fever. *The International Journal of Biochemistry & Cell Biology, 37*(8), 1560–1566.

Centers for Disease Control and Prevention. (2009). *Biosafety in microbiological and biomedical laboratories (BMBL)* (5th ed.). Washington, DC: U.S. Government Printing Office. Retrieved from http://www.cdc.gov/biosafety/publications/bmbl5/BMBL5_sect_IV.pdf

Centers for Disease Control and Prevention. (2013). *Viral Special Pathogens Branch (VSPB)*. Retrieved from http://www.cdc.gov/ncezid/dhcpp/vspb/index.html

Centers for Disease Control and Prevention. (2015a). *2014 Ebola outbreak in West Africa—Case counts*. Retrieved from http://www.cdc.gov/vhf/ebola/outbreaks/2014-west-africa/case-counts.html

Centers for Disease Control and Prevention. (2015b). *Outbreaks chronology: Ebola virus disease*. Retrieved from http://www.cdc.gov/vhf/ebola/outbreaks/history/chronology.html

Easton, A. J., & Pringle, C. R. (2011). Order *Mononegavirales*. In A. M. Q. King, M. J. Adams, E. B. Carstens, & E. J. Lefkowitz, *Virus Taxonomy—Ninth Report of the International Committee on Taxonomy of Viruses* (pp. 653–657). London, UK: Elsevier/Academic Press.

Feldmann, H., & Kiley, M. P. (1998). Classification, structure, and replication of filoviruses. *Current Topics in Microbiology and Immunology, 235*, 1–21.

Feldmann, H., Nichol, S. T., Klenk, H. D., Peters, C. J., & Sanchez, A. (1994). Characterization of filoviruses based on differences in structure and antigenicity of the virion glycoprotein. *Virology, 199*(2), 469–473.

Geisbert, T. W., & Jahrling, P. B. (1995). Differentiation of filoviruses by electron microscopy. *Virus Research, 39*(2), 129–150.

Gire, S. K., Goba A., Andersen, K. G., Sealfon, R. S., Park, D. J., Kanneh, L., … Sabeti, P. C. (2014). Genomic surveillance elucidates Ebola virus origin and transmission during the 2014 outbreak. *Science, 345*(6202), 1369–1372.

Harty, R. N., Brown, M. E., Wang, G., Huibregtse, J., & Hayes, F. P. (2000). A PPxY motif within the VP40 protein of Ebola virus interacts physically and functionally with a ubiquitin ligase: Implications for filovirus budding. *Proceedings of the National Academy of Sciences of the United States of America, 97*(25), 13871–13876.

Kuhn, J. H., Becker, S., Ebihara, H., Geisbert, T. W., Johnson, K. M., Kawaoka, Y., … Jahrling, P. B. (2010). Proposal for a revised taxonomy of the family *Filoviridae*: Classification, names of taxa and viruses, and virus abbreviations. *Archives of Virology, 155*(12), 2083–2103.

Licata, J. M., Simpson-Holley, M., Wright, N. T., Han, Z., Paragas, J., & Harty, R. N. (2003). Overlapping motifs (PTAP and PPEY) within the Ebola virus VP40 protein function independently as late budding domains: Involvement of host proteins TSG101 and VPS-4. *Journal of Virology*, *77*(3), 1812–1819.

Marí Saéz, A., Weiss, S., Nowak, K., Lapeyre, V., Zimmermann, F., Düx, A., ... Leendertz, F. H. (2014), Investigating the zoonotic origin of the West African Ebola epidemic. *EMBO Molecular Medicine*, *7*(1), 17–23.

National Public Radio. (2015, January). Death becomes disturbingly routine: The diary of an Ebola doctor. *Goats and Soda: Stories of Life in a Modern World* [Audio broadcast]. http://www.npr.org/blogs/goatsandsoda/2015/01/11/376362000/death-becomes-disturbingly-routine-the-diary-of-an-ebola-doctor

Negredo, A., Palacios, G., Vázquez-Morón, S., et al. (2011). Discovery of an ebolavirus-like filovirus in Europe. *PLoS Pathogens*, *7*(10), e1002304.

Okumura, A., Pitha, P. M., & Harty, R. N. (2008). ISG15 inhibits Ebola VP40 VLP budding in an L-domain-dependent manner by blocking Nedd4 ligase activity. *Proceedings of the National Academy of Sciences of the United States of America, 105*(10), 3974–3979.

Peterson, A. T., Bauer, J. T., & Mills, J. N. (2004). Ecologic and geographic distribution of filovirus disease. *Emerging Infectious Diseases*, *10*(1), 40–47.

Pourrut, X., Kumulungui, B., Wittmann, T., Moussavou, G., Délicat, A., Yaba, P., ... Leroy, E. M. (2005). The natural history of Ebola virus in Africa. *Microbes and Infection*, *7*(7), 1005–1014.

Pourrut, X., Souris, M., Towner, J. S., Rollin, P. E., Nichol, S. T., Gonzalez, J. P., & Leroy, E. (2009). Large serological survey showing cocirculation of Ebola and Marburg viruses in Gabonese bat populations, and a high seroprevalence of both viruses in *Rousettus aegyptiacus. BMC Infectious Diseases*, *9*(1), 159.

Swanepoel, R., Leman, P. A., Burt, F. J., Zachariades, N. A., Braack, L. E. O., & Ksiazek, T. G. (1996). Experimental inoculation of plants and animals with Ebola virus. *Emerging Infectious Diseases*, *2*, 321–325.

Taylor, D. J., Leach, R. W., & Bruenn, J. (2010). Filoviruses are ancient and integrated into mammalian genomes. *BMC Evolutionary Biology*, *10*, 193.

Turell, M. J., Bressler, D. S., & Rossi, C. A. (1996). Lack of virus replication in arthropods after intrathoracic inoculation of Ebola Reston virus. *American Journal of Tropical Medicine and Hygiene*, *55*, 89–90.

University of Texas Medical Branch at Galveston. (2013, May 2). Ebola's secret weapon revealed. *ScienceDaily*. Retrieved from http://www.sciencedaily.com/releases/2013/05/130502192226.htm

Wamala, J. F., Lukwago, L., Malimbo, M., Nguku, P., Yoti, Z., Musenero, M., ... Okware, S. I. (2010). Ebola hemorrhagic fever associated with novel virus strain, Uganda, 2007–2008. Emerging Infectious Diseases, 16(7). Retrieved from http://wwwnc.cdc.gov/eid/article/16/7/09-1525

Chapter 3

Brouseau, L., & Jones, R. (2014). *Ebola can be transmitted via infectious aerosol particles: Health workers need respirators, not masks*. Retrieved from http://www.globalresearch.ca/ebola-can-be-transmitted-via-infectious-aerosol-particles-health-workers-need-respirators-not-masks/5408022

Centers for Disease Control and Prevention. (2015, April 15). *Signs and symptoms*. Retrieved from: http://www.cdc.gov/vhf/ebola/symptoms

Ebola Response Team. (2014). Ebola virus disease in West Africa: The first nine months. *New England Journal of Medicine*, *371*, 1484–1495.

Grobbee, D. E., & Hoes, A. H. (2015). *Clinical epidemiology* (2nd ed.). Burlington, MA: Jones & Bartlett Learning.

Haas, C. (2014, October 14). On the quarantine period for Ebola virus. *PLOS Current Outbreaks*. Retrieved from http://currents.plos.org/outbreaks/article/on-the-quarantine-period-for-ebola-virus/

Hill, A. B. (1965). The environment and disease: Association or causation? *Proceedings of the Royal Society of Medicine*, *58*, 295.

Kobinger, G. P., Leung, A., Neufeld, J., Richardson, J. S., Falzarano, D., Smith, G., ... Weingartl, H. M. (2011). Replication, pathogenicity, shedding, and transmission of Zaire ebolavirus in pigs. *Journal of Infectious Diseases*, *204*, 200–208.

Rainisch, G., Shankar, M., Wellman, M., Merlin, T., & Meltzer, M. I. (2015). Regional spread of Ebola virus in West Africa. *Emerging Infectious Diseases*, *21*, 3. Retrieved from http://wwwnc.cdc.gov/eid/article/21/3/14-1845_article

Snow, J. (1849). On the pathology and mode of transmission of cholera. *London Medical Gazette*, *13*, 745–752.

Weingartl, H. M., Embury-Hyatt, C., Nfon, C., Leung, A., Smith, G., & Kobinger, G. (2012). Transmission of Ebola virus from pigs to non-human primates. *Nature Scientific Reports*, *2*, 811. doi:10.1038/srep00811.

World Health Organization. (2015). *Ebola*. Retrieved from http://apps.who.int/ebola/

Chapter 4

Ikegami, T., Niikura, M., Saijo, M., Miranda, M. E., Calaor, A. B., Hernandez, M., ... Morikawa, S. (2003). Antigen capture enzyme-linked immunosorbent assay for specific detection of Reston Ebola virus nucleoprotein. *Clinical and Diagnostic Laboratory Immunology, 10*(4), 552–557.

Iversen, P. L., Warren, T. K., Wells, J. B., Garza, N. L., Mourich, D. V., Welch, L. S., ... Bavari, S. (2012). Discovery and early development of AVI-7537 and AVI-7288 for the treatment of Ebola virus and Marburg virus infections. *Viruses*, *4*, 2806–2830.

Jones, D. (2009). Teaming up to tackle RNAi delivery challenge. *Nature Reviews Drug Discovery*, *8*(7), 525–526.

Kroll, D. (2014). *Why everyone has a stake in the Chimerix drug offered to Josh Hardy*. Retrieved from http://www.forbes.com/sites/davidkroll/2014/03/12/why-everyone-has-a-stake-in-the-chimerix-drug-offered-to-josh-hardy-cmx001-brincidofovir/

Kroll, D. (2015). *Chimerix ends brincidofovir Ebola trials to focus on adenovirus and CMV*. Retrieved from http://www.forbes.com/sites/davidkroll/2015/01/31/chimerix-ends-brincidofovir-ebola-trials-to-focus-on-adenovirus-and-cmv/

Leroy, E. M., Baize, S., Lu, C. Y., McCormick, J. B., Georges, A. J., Georges-Courbot, M. C., ... Fisher-Hoch, S. P. (2000). Diagnosis of Ebola haemorrhagic fever by RT-PCR in an epidemic setting. *Journal of Medical Virology*, *60*, 463–467.

Martin, P., Laupland, K. B., Frost, E. H., & Valiquette, L. (2015). Laboratory diagnosis of Ebola virus disease. *Intensive Care Medicine*, *41*(5), 895–898.

Qiu, X., Wong, G., Audet, J., Bello, A., Fernando, L., Alimonti, J. B., ... Kobinger, G. P. (2014). Reversion of advanced Ebola virus disease in nonhuman primates with ZMapp. *Nature*, *514*(7520), 47–53.

Samb, S., & Lewis, D. (2015, February 6). *Guinea to expand use of experimental anti-Ebola drugs*. Retrieved from http://www.reuters.com/article/2015/02/07/us-health-ebola-guinea-idUSKBN0LB0Q620150207

Warren, Travis K., Wells, J., Panchal, R. G., Stuthman, K. S., Garza, N. L., Van Tongeren, S. A., ... Bavari, S. (2014). Protection against filovirus diseases by a novel broad-spectrum nucleoside analogue BCX4430. *Nature*, *508*(7496), 402–405.

World Health Organization. (2014a). *Media briefing on the outcomes of the WHO/FIND meeting on diagnostic tests and Ebola control*. Retrieved from http://www.who.int/medicines/ebola-treatment/meetings/diagnostics-meeting-outcomes/en/

World Health Organization. (2014b). *WHO consultation on potential Ebola therapies and vaccines*. Retrieved from http://www.who.int/csr/resources/publications/ebola/ebola-therapies/en/

World Health Organization. (2015). Emergency use assessment and listing (EUAL) for Ebola virus disease (IVDs). Retrieved from http://www.who.int/diagnostics_laboratory/procurement/purchasing/en/

▶▶ Chapter 5

American Public Health Association. (1939). Mary Mallon (Typhoid Mary). *American Journal of Public Health/ Nation's Health*, *29*, 66–68.

Assessment Capacities Project. (2015). *Ebola outbreak in West Africa: Lessons learned from quarantine*. Retrieved from http://reliefweb.int/report/sierra-leone/ebola-outbreak-west-africa-lessons-learned-quarantine-sierra-leone-and-liberia

BBC News. (2014, April 1). *Ebola outbreak in Guinea "limited geographically"—WHO*. Retrieved from http://www.bbc.com/news/world-africa-26838885

BBC News Africa. (2014, October 17). *Ebola crisis: WHO accused of "Failure" in Ebola response*. Retrieved from http://www.bbc.com/news/world-africa-29668603

Centers for Disease Control and Prevention. (2014). *Detailed Emergency Medical Services (EMS) checklist for Ebola preparedness*. Retrieved from http://www.cdc.gov/vhf/ebola/pdf/ems-checklist-ebola-preparedness.pdf

Centers for Disease Control and Prevention. (2015). *Ebola*. Retrieved from http://www.cdc.gov/vhf/ebola/

Center for Infectious Disease Research and Policy. (2015). *Fast-track development of Ebola vaccines: Principles and target product criteria*. Retrieved from http://www.cidrap.umn.edu/sites/default/files/public/downloads/wellcome_trust-cidrap_ebola_vaccine_team_b_interim_report-final.pdf

Diallo, B. (2015). *Guinea closes border with Sierra Leone as it ramps up efforts to end Ebola*. Retrieved from http://www.usnews.com/news/world/articles/2015/03/30/guinea-shuts-border-with-sierra-leone-in-effort-to-end-ebola

HealthMap. (2014). *2014 Ebola outbreaks*. Retrieved from http://www.healthmap.org/ebola

Kuhnhenn, J. (2014, October 17). President Obama appoints Ebola "Czar." *AP News*.

McSpadden, K. (2015, April 20). WHO has acknowledged the failings of its Ebola response. *Time*. Retrieved from http://time.com/3827810/who-ebola-crisis-response-failure/

Médecins sans Frontières. (2014a). *Ebola: International response slow and uneven*. Retrieved from http://www.doctors withoutborders.org/article/ebola-international-response-slow-and-uneven

Médecins sans Frontières. (2014b). *Ebola: The failures of the international outbreak response*. Retrieved from http://www.msf.org/article/ebola-failures-international-outbreak-response

National Nurses United. (2014). *Press release: Nurses call on Obama to direct hospitals to follow highest standards for beating Ebola*. Retrieved from http://www.nationalnursesunited.org/press/entry/nurses-call-on-obama-to-direct-hospitals-to-follow-highest-standards-for-be/

Newcombe, T. (2014). Social media: Big lessons from the Boston Marathon bombings. *Emergency Management*. Retrieved from http://www.emergencymgmt.com/training/Social-Media-Lessons-Boston-Marathon-Bombing.html

Petrecca, L. (2013). After bombings, social media informs (and misinforms). *USA Today*. Retrieved from http://www.usatoday.com/story/news/2013/04/23/social-media-boston-marathon-bombings/2106701/

Sun, L. H., Dennis, B., Bernstein, L., & Achenbach, J. (2014, October 4). Out of control: How the world's health organizations failed to stop the Ebola disaster. *The Washington Post*. Retrieved from http://www.washingtonpost.com/sf/national/2014/10/04/how-ebola-sped-out-of-control/

U.S. Food and Drug Administration. (2014, November 6). *Inside clinical trials: Testing medical products in people*. Retrieved from http://www.fda.gov/Drugs/ResourcesForYou/Consumers/ucm143531.htm

World Bank. (2015). *Ebola: Most African countries avoid major economic loss but impact on Guinea, Liberia, Sierra Leone remains crippling*. Press release. Retrieved from https://www.worldbank.org/en/news/press-release/2015/01/20/ebola-most-african-countries-avoid-major-economic-loss-but-impact-on-guinea-liberia-sierra-leone-remains-crippling

World Food Programme. (2014). *Ebola leaves hundreds of thousands facing hunger in three worst-hit countries*. Retrieved from https://www.wfp.org/news/news-release/ebola-leaves-hundreds-thousands-facing-hunger-three-worst-hit-countries

World Health Organization. (2014a). *Drug resistance: Draft global action plan on antimicrobial resistance*. Retrieved from http://www.who.int/drugresistance/global_action_plan/en/

World Health Organization. (2014b). *Ebola outbreak in West Africa declared a public health emergency of international concern*. Retrieved from http://www.euro.who.int/en/health-topics/emergencies/pages/news/news/2014/08/ebola-outbreak-in-west-africa-declared-a-public-health-emergency-of-international-concern

World Health Organization. (2015). *Essential medicines and health products: Ebola vaccines, therapies, and diagnostics*. Retrieved from http://www.who.int/medicines/emp_ebola_section/en/

▶▶ Chapter 6

Bausch, D. G., Towner, J. S., Dowell, S. F., Kaducu, F., Lukwiya, M., Sanchez, A., … Rollin, P. E. (2007). Assessment of the risk of Ebola virus transmission from bodily fluids and fomites. *Journal of Infectious Diseases, 196*(Suppl 2), S142–S147.

Centers for Disease Control and Prevention. (2012). *Rodents in the United States that carry Hantavirus*. Retrieved from http://www.cdc.gov/hantavirus/rodents/index.html

Centers for Disease Control and Prevention. (2013). *Viral hemorrhagic fevers*. Retrieved from http://www.cdc.gov/ncidod/dvrd/spb/mnpages/dispages/vhf.htm

Centers for Disease Control and Prevention. (2014). *Infection control for viral haemorrhagic fevers in the African health care setting*. Retrieved from http://www.cdc.gov/vhf/abroad/vhf-manual.html

Centers for Disease Control and Prevention. (n.d.). *Hendra virus disease and Nipah virus encephalitis: Fact sheet*. Retrieved from http://www.cdc.gov/ncidod/dvrd/spb/mnpages/dispages/Fact_Sheets/Hendra_Nipah_Fact_Sheet.pdf

Centers for Disease Control and Prevention. (2015). Morbidity and Mortality Weekly Report (MMWR): Possible Sexual Transmission of Ebola Virus — Liberia. Retrieved from http://www.cdc.gov/mmwr/preview/mmwrhtml/mm6417a6.htm

Gryseels, S., Rieger, T., Oestereich, L., Cuypers, B., Borremans, B., Makundi, R., … Goüy de Bellocq, J. (2015). Gairo virus, a novel arenavirus of the widespread *Mastomys natalensis*: Genetically divergent, but ecologically similar to Lassa and Morogoro viruses. *Virology, 476*, 249–256.

Metabiota. (2014). Case study: *Big data on little animals: The Global Animal Information Network System (GAINS)*. Retrieved from http://www.metabiota.com/media/CaseStudy-GAINS1.pdf

Pernet, O., Schneider, B. S., Beaty, S. M., LeBreton, M., Yun, T. E., Park, A., … Lee, B. (2014). Evidence for henipavirus spillover into human populations in Africa. *Nature Communications, 5*, 5342.

Phillip, A. (2014, September 5). How Ebola is stealing attention from illnesses that kill more people. *Washington Post*. Retrieved from http://www.washingtonpost.com/news/to-your-health/wp/2014/09/05/how-ebola-the-kardashian-of-diseases-is-stealing-attention-from-illnesses-that-kill-more-people/

Rogstad, K. E., & Tunbridge, A. (2015). Ebola virus as a sexually transmitted infection. *Current Opinion in Infectious Diseases, 28*(1), 83–85.

U.S. Department of the Interior, U.S. Geological Survey. (2015). *Disease Maps 2014*. Retrieved from http://diseasemaps.usgs.gov/index.html

Wolfe, N. (2014, October 14). No more Ebola Whac-A-Mole. *Wall Street Journal*. Retrieved from http://www.wsj.com/articles/nathan-wolfe-no-more-ebola-whac-a-mole-1413241442

World Health Organization. (2011). *Final report of the IHR Review Committee published*. Retrieved from http://www.euro.who.int/en/health-topics/communicable-diseases/influenza/news/news/2011/05/final-report-of-the-ihr-review-committee-published

World Health Organization. (2013). *Crimean-Congohaemorrhagic fever*. Retrieved from http://www.who.int/mediacentre/factsheets/fs208/en/

World Health Organization. (2014). *Yellow fever*. Retrieved from http://www.who.int/mediacentre/factsheets/fs100/en/

World Health Organization. (2015). *One year into the Ebola epidemic: A deadly, tenacious, and unforgiving virus*. Retrieved from http://www.who.int/csr/disease/ebola/one-year-report/ebola-report-1-year.pdf?ua=1

Zipprich, J., Hacker, J. K., Murray, E. L., Xia, D., Harriman, K., & Glaser, C. (2014). Notes from the field: Measles—California, January 1–April 18, 2014. *Morbidity and Mortality Weekly Report, 63*(16), 362–363.

Index

Note: Page numbers followed by *f* or *t* indicate material in figures or tables respectively.